# Using *MultiSIM*:
# Digital Electronics

# Using *MultiSIM*:
# Digital Electronics

## John Reeder
Merced College
Merced, CA

**DELMAR**

**THOMSON LEARNING** ™

Australia   Canada   Mexico   Singapore   Spain   United Kingdom   United States

**DELMAR**

™

**THOMSON LEARNING**

Using *MultiSIM*: Digital Electronics
John Reeder

**Business Unit Director:**
Alar Elken

**Executive Editor:**
Sandy Clark

**Senior Acquisitions Editor:**
Gregory L. Clayton

**Developmental Editor:**
Michelle Ruelos Cannistraci

**Executive Marketing Manager:**
Maura Theriault

**Marketing Coordinator:**
Karen Smith

**Executive Production Manager:**
Mary Ellen Black

**Production Manager:**
Larry Main

**Senior Project Editor:**
Christopher Chien

**Art/Design Coordinator:**
David Arsenault

**Editorial Assistant:**
Jennifer M. Luck

For permission to use material from the text or product,
contact us by
Tel.   (800) 730-2214
Fax   (800) 730-2215
www.thomsonrights.com

ISBN: 07668-12693

**NOTICE TO THE READER**

# Contents

# Preface

## Approach

This workbook is designed to teach the student how to virtually measure, operate, and troubleshoot digital electronic circuits created within the *MultiSIM 2001*® software environment and to reinforce digital theory learned in the classroom. The computer and the computer monitor become the electronics workbench. Students using this manual must have *MultiSIM 2001* installed on their computers to be able to operate the provided software projects. These software projects, along with the textbook edition of *Using MultiSIM,* are stored on the CD that accompanies this text.

One advantage to the virtual laboratory approach to electronics is the low cost of the software package in comparison to the expenses required to establish an electronics laboratory with all of the necessary test equipment and the related facility costs. Virtual electronic circuits can be easily modified on the monitor screen, and circuit analysis is also easily obtained as circuits are modified. Simultaneously, this software can be used as an additional tool to help the electronics student learn more about digital theory.

Circuit troubleshooting is an integral component of the software package. It is relatively easy for the instructor or textbook author to install faults, such as shorts, leakage, and opens, into the circuit for the digital electronics student to locate. The troubleshooting exercises will provide the student with the confidence and skills necessary to troubleshoot circuits constructed on the electronics laboratory workbench.

One additional technique used in this book to reinforce learning is a construction project at the end of many of the activities. These projects will help the student work through the connection requirements of complex integrated circuits.

## System Requirements

Pentium 166 or greater PC
Windows 95/98/NT
32MB RAM (64MB RAM recommended)
**100-250 MB hard disk space (min.)**
CD-ROM drive
$800 \times 600$ minimum screen resolution

# Organization

Each chapter in this workbook attempts to sequentially follow the material found in most textbooks teaching digital theory. **Activity** sections, located in each chapter, divide the overall subject of the chapter into smaller blocks. The individual software projects are related to subtopics within the larger topic.

Within each activity, **circuit files** progressively provide individual projects related to the subject material of the chapter. Many related subjects are touched upon as progress is made through the projects. Each activity has one or more circuit files containing component and connection faults, which are **troubleshooting** problem/s. Finally, most of the activities contain a construction circuit requiring the student to construct and test a circuit related to the subject of the activity.

# Circuit Files

The CD that accompanies the text contains all of the circuit files in this book. They are pre-built and ready to be used with *MultiSIM 2001*.

Circuit files follow the DOS system of nomenclature. There is a maximum of eight digits in each circuit name, but unlike previous versions of *Electronics Workbench®*, the circuit file name is not followed by a DOS extension. The first two numbers of the circuit file name represent the chapter in the book and are followed by a hyphen. After the hyphen, the next two numbers represent the activity within the chapter. The final letter, in most cases, represents the sequence of events within an activity. The DOS extension for *MultiSIM* files is .msm.

# Concerning Instrumentation

Test instruments are accessed through the **Instruments** button on the Design Bar above the circuit workspace. A left-click on this button will cause the instruments toolbar to appear at the right side of the screen. You will use the DMM, the oscilloscope, and the function generator with this workbook. The DMM, the oscilloscope, and the function generator are common instruments found on the electronics test bench. One new feature of *MultiSIM 2001* is the ability to place more than one piece of the same type of test instrument, such as two oscilloscopes, on the workspace simultaneously.

# Additional Resources

Additional support for electronics instructors is provided on the publisher's web site at **www.electronictech.com**. Answers to questions and problems in the text will be provided in a password-protected location accessible only to instructors.

*MultiSIM* can be purchased through your college bookstore or through **www.electronictech.com**. Interactive Image Technologies, the producers of *Electronics Workbench®*, *MultiSIM 2001*, can be contacted at (416) 977-5550 for sales or technical support. Their web page is:

**www.electronicsworkbench.com**.

## About the Author

John Reeder, A.A., B.A., M.S., is an electronics instructor at Merced College and also taught high school electronics for nine years. Prior to becoming a teacher, he worked for 28 years in the electronics and electrical industries as a technician, electrician, and engineer. He has been using *Electronics Workbench®* and *MultiSIM* for many years as a supplement to help students gain a better understanding of their electronics material. At Merced College, he has developed a course curriculum using *EWB* and *MultiSIM* as the software core of the overall electronics program. These electronics software classes at Merced College are corequisite requirements for all foundational classes in the electronics program.

## Acknowledgments

I would like to express my grateful appreciation to the publishing team at Delmar Thomson Learning for patiently working with me and helping me over the hurdles of a second book. The members of this team are Greg Clayton, Michelle Ruelos Cannistraci, Larry Main, Chris Chien, David Arsenault, Jennifer Luck, and Jennifer Thompson.

I would also like to express my grateful appreciation to the software team at Electronics Workbench, Scott Duncan, Luis Alves, Roman Bysh, and Kevin Braham, who helped me through the difficult moments when the software wasn't responding to my efforts. All worked out well and the result is this finished product.

I want to thank my partners in electronics instruction at Merced College, Eugen Constantinescu and Bill Walls, who reviewed the initial draft of the book and software projects. Their review of the initial draft and software is greatly appreciated.

I also want to thank my student, Sharol Stang, who reviewed the book from the student vantage point and also my electronics students, who provided the original impetus to get me started in writing and developing these files for their use. Their enthusiasm over the book and the *MultiSIM* projects is wonderful.

Last, but not least, I want to acknowledge the most important member of my team here in Merced, my wife Barbara, who put up with the long hours spent on the computer over a summer break, winter break, weekends, and every other spare moment. She always provided the right touch (coffee and chocolate) at the right moments and kept me going at critical points in the writing process.

While writing the book, I referred to many digital texts published by various authors and publishers. Among my primary sources of reference was the excellent text authored by Robert Dueck, *Digital Design with CPLD Applications and VHDL* and the text authored by James Bignell and Robert Donovan, *Digital Electronics,* both of which are published by Delmar Thomson Learning. I with to especially give my thanks to Robert Dueck for his kind permission to use his definitions regarding the difference between latch and flip-flop circuits.

# 1. Introduction to Digital Concepts

**References**

*MultiSIM* 2001

*MultiSIM* 2001 User's Guide

**Objectives**   After completing this chapter, you should be able to:

- Understand and use the Binary number system.
- Use timing diagrams of digital circuits.
- Understand and use the term BIT (binary digit).
- Differentiate between analog and digital.
- Calculate period and frequency of digital waveforms on an oscilloscope.
- Interpret a digital waveform in terms of highs and lows (1s and 0s).
- Understand the basic concepts of Boolean Algebra.

## Introduction

### Number Systems

Digital logic circuits operate in the binary mode where the inputs have two possible states, a binary 1 represented normally by a "HIGH" voltage and a binary 0 represented normally by a "LOW" voltage. Digital logic can be developed with the use of logic symbols and the Boolean algebra equations that define the logic. Logic gates are redundant electronic circuits that combine digital logic signals in specific patterns according to the dictates of the related Boolean algebra equation.

Other number systems used in digital electronics include the **Octal** and **Hexadecimal** systems. The octal system is a number system with a base of 8, and the hexadecimal system is a number system with a base of 16. Both of these systems work well with the binary system with its base of 2, being powers of that base (i.e., $8 = 2^3$ and $16 = 2^4$).

### Boolean Algebra

In 1854, George Boole published a work entitled *An Investigation of the Laws of Thought, on Which are Founded the Mathematical Theories of Logic and Probabilities*. In this work, Boole developed a system of logical algebra known in the modern scientific world as **Boolean algebra**. After he completed his work, George Boole's concepts gathered dust until Claude Shannon used Boole's work to develop his master's thesis at Massachusetts Institute of Technology. His thesis was entitled "A Symbolic Analysis of Relay and Switching Circuits."

The reintroduction of Boole's logic, by Shannon, into modern technical thought and usage assumed the title of Boolean algebra. This method of logic greatly simplified the design, development, and application of complex telephone logic circuits and ultimately, the logic circuit requirements of the newly developing field of digital computers.

All functions, no matter how complex, that may be required in digital logic, can be created from three basic logic functions, AND, OR, and NOT. The basic logic gates that you will study in this text can be used to solve Boolean algebra equations. The algebraic logical variables are the inputs to the logic gates, and the logical functions and/or formulas are the outputs of these gates or combination of gates. The more complex gates such as the NAND and NOR can be thought of as combinations of ANDs, ORs, and NOTs.

# Activity 1.1: The Concept of a Digital Circuit

1. In the electronics world, there are basically two types of circuits, the analog circuit and the digital circuit. We are more familiar with analog circuits because most of the natural phenomena that we deal with in the physical world are analog in nature. They are always changing and have an infinite number of variables between any two points of measurement. In the realm of voltages, analog voltages can and do vary continuously at the inputs and outputs of these types of circuits. That is their nature.

2. On the other hand, digital circuits are generally referred to as circuits that deal only in highs and lows with discrete binary values. When discussing digital circuits, we use the terms **HIGH** and **LOW**. We are referring to voltages with only two distinct possibilities, a high voltage or a low voltage.

3. Open circuits file **d01-01**. Turn on the circuit and watch the lamps.

   Which bulb is steady state and which bulb is changing? X _____ is steady

   state, and X _____ is changing.

4. As you can see, X1 is in a digital type of circuit (direct current and a fixed voltage) in the HIGH condition and X2 is in an analog circuit (alternating current and an alternating voltage).

5. Many electronic systems use a combination of analog and digital circuits. The CD player is an example. The data that originates on the compact disk itself is digital in nature, but that data is usually analog in origin. The digital data is typically a binary representation of music that is analog,

the CD is the medium for storage only, and the purpose of the CD player is to restore the digital information to its original analog form for the listener. The music was recorded using an analog-to-digital converter (ADC) and the music is restored using a digital-to-analog converter (DAC).

## Activity 1.2: Binary Digits and Logic Levels

1.  Specific HIGHs and LOWs in a binary system are referred to as **bits**, an acronym coming from the term **binary digits**. The two digits in the binary number system are **1** (one) and **0** (zero). In a digital circuit, a binary 1 is usually referred to as a HIGH voltage and the 0 is usually represented as a LOW voltage. A common voltage representing a binary 1 is +5 VDC and 0 V usually represents a binary 0. Some electronic systems, for various reasons, use negative logic where HIGHs and LOWs have opposite meanings and some families of integrated circuits (ICs) use voltage levels other than 0 and +5 volts.

2.  The other factor regarding binary digits is in the realm of time. Many times HIGHs and LOWs are referred to in relationship to a period of time. Figure 1-1 displays a typical digital waveform on the screen of an oscilloscope. This display of a digital waveform has to be interpreted in relationship to time. The duration of one cycle of the waveform can be measured in time; each square on the scope face represents a calibrated period of time. In this case, each square is equal to 1 millisecond (.001 seconds) and one complete cycle of an individual waveform takes one square or 1 millisecond (the period of one square is referred to as the Timebase). Using the formula: Frequency = 1/Time (or Period), you can calculate the frequency of the digital waveform. The frequency of this

waveform is equal to _____ Hertz (Hz).

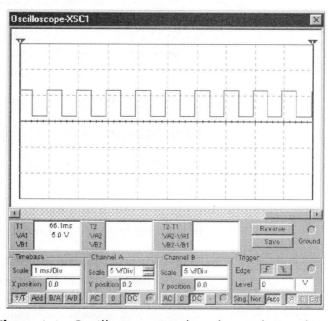

**Figure 1-1**  Oscilloscope Display of Digital Waveform

3. If we specify that the duration of one bit of binary data (in the case of this specific waveform) is one half of a square or 0.5 milliseconds in duration, then we can also state that each square contains one HIGH and one LOW. Also, we can state that the waveform is HIGH for 0.5 milliseconds (or 500 microseconds) and LOW for 500 microseconds (500 μseconds). This period of time is called **bit time**.

4. Figure 1-2 displays a digital pulse waveform with a 20% duty cycle (HIGH 20% of the time). If the timebase is 50 μseconds, what is the duration of one cycle of the waveform? One cycle of the waveform is

   _____ μseconds in duration. What is the frequency of the waveform? The

   frequency of the waveform is _____ kHz or _____ Hz (1 kHz = 1000 Hz).

5. The amount of time that one bit of digital data in a binary sequence occupies is defined as bit time. If the bit time for the waveform of Figure 1-2 is 10 μseconds and the time between bits is 90 μseconds,

   then we can state that this waveform is HIGH for _____ bit(s) and

   LOW for _____ bits.

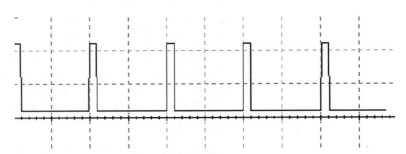

**Figure 1-2**   Digital Pulse Waveform

6. In a binary sequence of HIGH, HIGH, LOW, LOW, LOW, HIGH, LOW, HIGH, HIGH, and HIGH (read from left to right), we can state in binary that the message is (in the same order) 1100010111.

7. What would be the binary message for HIGH, LOW, LOW, LOW, HIGH, HIGH, LOW, HIGH? The binary message would be _____

   _____.

## Activity 1.3: Boolean Algebra—AND, OR, and NOT Functions

1. Boolean algebra has become the preferred method of specifying the logic of digital circuits. Boolean can be thought of as the shorthand of digital circuits, and the Boolean expression as the defining factor in circuit design. Boolean is a universal digital language understood by

programmers, engineers, technicians, and anyone else involved in digital circuitry.

2. All logic functions, no matter how complex, that are required in digital logic can be created from three basic logic functions, AND, OR, and NOT. The basic logic gates to be studied in the following chapters are examples of solutions to Boolean algebra equations. The Boolean algebra variables are the logic inputs to our logic gates and the logical function or formula is the output of the gate or of even more complex logic circuits. The more complex gates, such as the NAND and NOR, can be thought of as at the bottom level as various combinations of AND, OR, and NOT gates.

3. **AND Logic** can be referred to as 'all-or-nothing' logic. Using the terms $A$ and $B$ as inputs and the term $Y$ as an output, we can state, using AND logic, that inputs $A$ AND $B$ have to be HIGH for output $Y$ to be HIGH. This can be stated in terms of Boolean algebra as $A \cdot B = Y$ (the $\cdot$ represents the AND function).

4. Open circuits file **d01-03a**. In this circuit, switches A, B, and C are connected in an AND (series) configuration. Closing each switch with the related computer keyboard key represents a HIGH condition. What is the AND formula that will turn on the lamp? The AND formula that will

   turn on the lamp is _____.

5. **OR Logic** can be referred to as 'any-or-all' logic. Again, using the terms $A$ and $B$ as inputs and the term $Y$ as an output, we can state, using OR logic, that if input $A$ OR $B$ is HIGH (or both $A$ and $B$), then output $Y$ will be HIGH. This can be stated in terms of Boolean algebra as $A + B = Y$ (the $+$ represents the OR function).

6. Open circuits file **d01-03b**. In this circuit, switches A, B, and C are connected in a parallel OR configuration. Closing one of the switches represents a HIGH condition for that switch. What is the OR formula for this circuit that will turn on the lamp? The OR formula that will turn the

   lamp ON is _____.

7. **NOT Logic** can be referred to as 'inverted, negated, or complement' logic. The NOT function changes one logic level to the opposite logic level. The NOT function in a Boolean algebra formula is represented by an overbar over the individual input, such as $\overline{A}$, or over a whole function, such as $\overline{A + B + C} = Y$. In this second example, you would read the formula as the "quantity $A$ OR $B$ OR $C$ all NOTTED is equal to $Y$". Sometimes the apostrophe symbol is used to represent the NOT function such as $A'$ for NOT $A$ or $A$ prime. The overbar will be used in the text and the apostrophe will be used with MultiSIM circuit diagrams.

8. Open circuits file **d01-03c**. In this circuit, if switch A' is open, then lamp Y is ON, representing a HIGH. If switch A' is closed, then lamp Y

is OFF, representing a LOW. What is the Boolean equation for this circuit? The Boolean equation for this circuit is _____

_____.

## Activity 1.4: Troubleshooting Problems

1. Troubleshooting problems will be found in this text after every activity. There will be two basic types of troubleshooting problems.

   • One type of problem will be a connection problem where the circuit will be connected incorrectly. Many times there will be a statement regarding how to connect a certain type of integrated circuit and that clue will be the key to finding the connection problem.

   • The other type will be a component fault introduced by the instructor. This type of problem will consist of opens and shorts. If, for example, all of the inputs to a circuit are correct, then the problem might be an open output. Many times there will be shorted inputs or outputs. Switches and power sources can be open or shorted.

2. Open circuits file **d01-04a**. This circuit has a connection problem. Actuate the circuit and the switches. Did the lamp turn on, Yes _____ or No _____? It should not have turned on because of the connection problem. The wire from the bottom of the lamp is connected to junction B rather than junction A. Move the wire from junction B to junction A. Activate the circuit and close the switches. Did the lamp turn on, Yes _____ or No _____?

3. Open circuits file **d01-04b**. This circuit has a component problem. Actuate the circuit and the switch. Did the lamp turn on, Yes _____ or No _____? It should not have turned on because of the circuit problem. Use the Digital Multimeter (DMM) to troubleshoot the circuit. Measure the voltage output of the power source. What was the value of the output voltage? The output voltage was _____ V. Obviously, the problem is the lack of voltage from the power source. Replace V 1 with V 2. Does the circuit work now, Yes _____ or No _____?

4. Open circuits file **d01-04c**. This circuit has a component problem. Actuate the circuit and the switch. Did the lamp turn on, Yes _____ or No _____? It should not have turned on because of the circuit problem. Use the Digital Multimeter to troubleshoot the circuit. Measure the

voltage of the voltage source. The voltage of the voltage source is _____ V. Measure the voltage drop across the switch. When the switch is open, the voltage drop across the switch is _____ V, and when the switch is closed, the voltage drop across the switch is _____ V. Obviously, the switch is defective because the voltage drop should be 0 V when the switch is closed. Replace J 1 with J 2. Does the circuit work now, Yes _____ or No _____?

5. Open circuits file **d01-04d**. This circuit has a component problem. Actuate the circuit and the switch. Did the lamp turn on, Yes _____ or No _____? It should not have turned on because of the circuit problem. Use the Digital Multimeter to troubleshoot the circuit. Measure the voltage of the voltage source. The voltage of the voltage source is _____ V. Measure the voltage drop across the switch. When the switch is open, the voltage drop is _____ V, and when the switch is closed, the voltage drop is _____ V. The power source and the switch check okay, so that leaves the lamp. Measure the voltage drop across the lamp. When the switch is closed, the voltage drop across the lamp is _____ V. Obviously, the problem in this circuit is the lamp, there should be a voltage drop of 12 V across the lamp when the switch is closed. Measure the resistance of the lamp. The resistance of lamp X1 is _____ ohms. Replace lamp X1 with lamp X2. Reactivate the circuit. Did the lamp light as it should, Yes _____ or No _____?

## Activity 1.5: Circuit Construction

1. Circuit construction problems will be found in this text after every activity. You will be required to construct a circuit that is similar to the particular type of circuit that you just studied in the activity.

2. Open circuits file **d01-05a**. Using the components on the workspace, construct a circuit similar to Figure 1-3 on page 8. Activate the switches.

    Did lamps X1 and X2 turn on properly, Yes _____ or No _____?

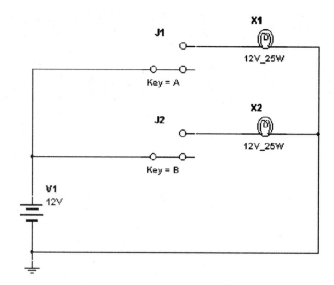

**Figure 1-3**   Construction Circuit for Step 2

3. Open circuits file **d01-05b**. Using the components on the workspace construct a circuit similar to Figure 1-4. Activate the circuit. Did the lamp

   light, Yes _____ or No _____?

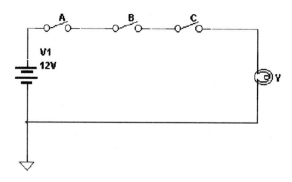

**Figure 1-4**   Construction Circuit for Step 3

# 2. Logic Gates and Combinational Circuits

---

*References*

*MultiSIM* 2001

*MultiSIM* 2001 User's Guide

---

**Objectives**  After completing this chapter, you should be able to:

- Use MultiSIM® circuits to perform basic logic operations.
- Determine truth table information for basic logic gates.
- Develop timing diagrams for basic logic gates.
- Use Boolean algebra to construct logic circuits.
- Use NAND and NOR gates to implement Boolean algebra equations.
- Construct logic circuits.
- Troubleshoot logic circuits.

## Introduction

Digital logic circuits (gates) operate in the binary mode where the inputs have only two possible states, a binary 1 represented normally by a HIGH voltage and a binary 0 represented normally by a LOW voltage. Digital logic can be developed with the use of logic symbols and the Boolean algebra equations that define the logic. Logic gates are redundant electronic circuits that combine digital logic signals in specific patterns according to the dictates of the related Boolean algebra equation. The basic logic gates that we will study in this chapter are:

- The **Inverter** or **NOT** gate
- The **OR** gate
- The **AND** gate
- The **NAND** gate
- The **NOR** gate

## Activity 2.1: The Inverter or NOT Gate

1. The basic **inverter** is a single-input gate with a single output. The output (Y) is the complement of the input (A). If the input signal is a binary 1,

---

then the output is a binary 0 and if the input is a binary 0, then the output is a binary 1. The Boolean logic of an inverter states that if A is 0, then Y is 1 and if A is 1, then Y is 0. Figure 2-1 shows several logic symbols for an inverter, one with a bubble (circle) at the input and the other with a bubble at the output. The bubble at the input represents an **active LOW** input and the absence of a bubble at the input indicates an **active HIGH** input. The Boolean term for the output is $\overline{A}$ (the same as Y) and is read A complement or A NOT (notice the line over the top of the letter A, this indicates the NOT function). Another name for an inverter is a **NOT** gate.

**Figure 2-1**   Inverter Gate Logic Symbols

2. When using TTL circuits, the HIGH condition (binary 1) is represented by +5 V and 0 V represents the LOW condition (binary 0). In reality, the IC input voltage for a HIGH condition can be from +2.0 V to +5 V and the input LOW condition can be from 0 V to +0.8 V. The HIGH state at the output can be from +2.4 V to +5 V and the LOW state at the output can be from 0 V to +0.4 V. Other families of ICs such as the CMOS family will have different input and output voltages. The simulated voltages that you measure in your MultiSIM circuits probably will display values other than the typical textbook statements of 0 V and +5 V for binary ones and zeros.

3. Open circuits file **d02-01a** and observe the input and output voltage readings of the inverter circuit. Notice that the complementary voltages change when you change the input with the **SPACE** bar. If the input is

   0 V, the output meter measures _____ V. If the input is 5 V, the output

   meter measures _____ V. It may take a while for the meters to settle down, be patient. The reading only needs to reflect one decimal point (for example, 4.4 V rather than 4.433 V). Go ahead and round off your answers.

4. One of the most common inverter ICs used in industry is the 7404 (TTL family) **hex inverter**. This circuit is called a hex inverter because there are six independent inverter circuits inside the 7404 dual-inline package (DIP). Figure 2-2 displays the pinout diagram of the 7404 circuit. Notice that the inputs are designated as A (1A, 2A, and so on) and the outputs are designated as Y (1Y, 2Y, and so on). For example, the circuit 1 input is pin 1 of the IC and the circuit 1 output is pin 2 of the IC.

5. Open circuits file **d02-01b**. In this circuit, the 7404 hex inverter circuit is used. The U1A output (pin 2) is connected to the input of the second circuit (U1B), the output of the second circuit is connected to the input

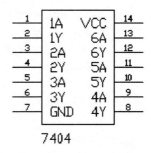

**Figure 2-2**  Pinout for a 7404 Hex Inverter

of the third circuit (U1C), and so on. Use the DMM and measure the outputs of each circuit and enter your data into Table 2-1. Also enter the condition of the voltage probes at each output (on or off). Data for the first section is already entered as an example and the DMM is connected for this initial measurement.

| Switch 1 Input Voltage | Inverter Outputs | | | | | |
|---|---|---|---|---|---|---|
| | Section U1A (Pin 2) | | Section U1B (Pin 4) | | Section U1C (Pin 6) | |
| | V Probe | Volts | V Probe | Volts | V Probe | Volts |
| LOW (0 V) | On | 5 V | | | | |
| HIGH (5 V) | Off | 0 V | | | | |
| | Inverter Outputs | | | | | |
| | Section U1D (Pin 8) | | Section U1E (Pin 10) | | Section U1F (Pin 12) | |
| | V Probe | Volts | V Probe | Volts | V Probe | Volts |
| LOW (0 V) | | | | | | |
| HIGH (5 V) | | | | | | |

**Table 2-1**  7404 Hex Inverter Data

6. Open circuits file **d02-01c**. Use the components on the workspace to make a circuit similar to the circuit of Step 4. Make the same measurements as in Step 4 and enter your measurements and observations into Table 2-2.

| Switch 1 Input Voltage | Inverter Section | | | | | |
|---|---|---|---|---|---|---|
| | Section U1A (Pin 2) | | Section U1B (Pin 4) | | Section U1C (Pin 6) | |
| | V Probe | Volts | V Probe | Volts | V Probe | Volts |
| LOW (0 V) | | | | | | |
| HIGH (5 V) | | | | | | |
| | Inverter Section | | | | | |
| | Section 1D (Pin 8) | | Section 1E (Pin 10) | | Section 1F (Pin 12) | |
| | V Probe | Volts | V Probe | Volts | V Probe | Volts |
| LOW (0 V) | | | | | | |
| HIGH (5 V) | | | | | | |

**Table 2-2**   7404 Hex Inverter Data (Student Circuit)

## ● *Troubleshooting Problems:*

7. When working with troubleshooting problems in digital circuits, the most common defects that can be traced directly to integrated circuits are usually shorted or open conditions located at the inputs and/or outputs of the circuits. Troubleshooting these types of defects can be accomplished with the DMM or the voltage probe that is provided with most of the troubleshooting problems throughout the text.

   a. If the circuit defect is a shorted condition, then the short is reflected to all terminals connected to the short. This reflection of a shorted condition makes isolation of the shorted location somewhat difficult. The various terminals have to be disconnected one at a time until the short goes away if the short is caused by an integrated circuit. If the shorted condition is caused by another component such as the printed circuit board, then the first method of defect isolation is a visual inspection. If the visual inspection fails to locate the short, then more sophisticated methods are necessary which are beyond the scope of this textbook. A step-by-step isolation procedure is always the best method to locate a fault of any kind.

   b. If the problem is an open condition, then the problem is a little easier to isolate. The desired signal starts from one location and fails to reach another location. This type of problem can be caused by broken wires, broken printed circuit board traces, bent (under) pins on integrated circuits preventing normal socket connections, a faulty IC, or other possibilities that can be isolated visually or by means of test equipment.

c. Many times shorted conditions or missing signals can be the integrated circuits themselves. First make sure that an open or short external to the output terminals is not the source of the problem. Then check all input terminals for the proper signals. If the problem seems to be internal to the integrated circuit, then replacement of the suspect integrated circuit is usually the next step. This procedure can be difficult if the IC is soldered to a printed circuit board (PCB).

8. Open circuits file **d02-01d**. One of these circuits has an open input. Use the DMM and the voltage probes to find the location of the open. Then disconnect wires between terminals to isolate the problems to the output of one circuit or the input of another. The open condition is located at the

(input/output) of _____. Notice that the A key on the keyboard actuates the switch (rather than the space bar).

9. Open circuits file **d02-01e**. When this circuit is turned on, it appears to be a dead circuit. This type of condition can frequently be traced to a power supply problem (hint). Use the DMM to locate the problem.

The reason for the dead circuit is_____

_____

_____.

10. Open circuits file **d02-01f**. The problem with this circuit is that the output does not reflect a change in the input voltage (starting with U10). The fault condition, in this case no change in the lamps after U10, has to be isolated. Use the DMM to locate the location of the *no signal* condition.

The *no signal* condition is located at the output of U____ and the reason

for the *no signal* condition is _____

_____.

11. Open circuits file **d02-01g**. This circuit has an internal problem in the integrated circuit with a *no signal* condition at the final output. The *no signal* condition, in this case is a low output and there is no change in the output when the input changes. Use the DMM to locate the problem.

The fault is located at the input/output of U_____ and it is

_____

(what is wrong with the IC).

● *Circuit Construction:*

12. Open circuits file **d02-01h**. Using the components and circuit on the workspace, construct an inverter circuit that will turn on X0 and X2 when the switch (spacebar) is up and turn on X1 and X3 when the switch is down. Use all of the components.

## Activity 2.2: The OR Gate

1. The **OR** gate is a circuit that places a binary one (HIGH) at its output if any one of its inputs are HIGH. If all of its inputs are binary zeros or LOW, then the output will be LOW also. Figure 2-3 shows the logic symbol for a two-input OR gate and also a **truth table** to demonstrate all possible input and output conditions. The Boolean equation for this OR gate is $A + B = Y$. The + symbol indicates the OR function.

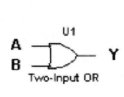

| Truth Table for Two-Input OR Gate | | |
|---|---|---|
| B | A | Y |
| 0 | 0 | 0 |
| 0 | 1 | 1 |
| 1 | 0 | 1 |
| 1 | 1 | 1 |

**Figure 2-3**   Two-Input OR Gate Symbol and Truth Table

2. Open circuits file **d02-02a**. Actuate the A input and B input switches according to the truth table and fill in the results (Y) in Table 2-3. The voltage probe indicates the HIGH or LOW condition of output Y. The

Boolean equation for this circuit is _____

_____.

| OR Truth Table | | |
|---|---|---|
| B | A | Y |
| 0 | 0 | |
| 0 | 1 | |
| 1 | 0 | |
| 1 | 1 | |

**Table 2-3**   OR Gate Truth Table

3. The pinout of a typical 74 series OR gate is shown in Figure 2-4. This OR gate is a 7432 Quad two-input integrated circuit. The IC contains four individual OR gates with each gate having two inputs.

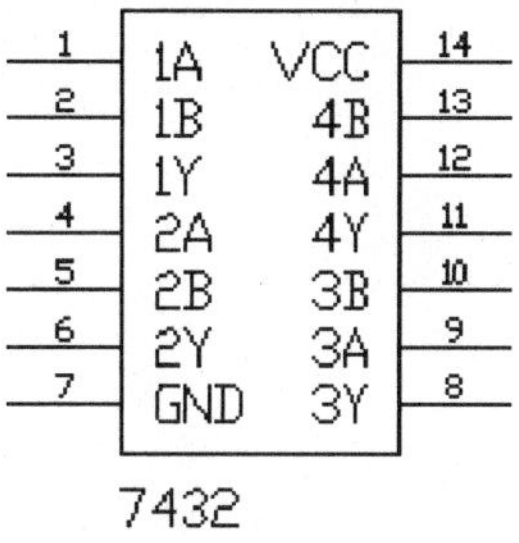

```
     1 ┌─────────────┐ 14
  ─────┤ 1A      VCC ├─────
     2 │             │ 13
  ─────┤ 1B       4B ├─────
     3 │             │ 12
  ─────┤ 1Y       4A ├─────
     4 │             │ 11
  ─────┤ 2A       4Y ├─────
     5 │             │ 10
  ─────┤ 2B       3B ├─────
     6 │             │ 9
  ─────┤ 2Y       3A ├─────
     7 │             │ 8
  ─────┤ GND      3Y ├─────
       └─────────────┘
            7432
```

**Figure 2-4**   Pinout for a 7432 OR Gate

4. Open circuits file **d02-02b**. In this circuit, only the first OR gate is being used and has a voltage probe connected to output 1Y to indicate the output condition of the gate. Actuate the input switches, A and B, and enter your data into Table 2-4.

| 7432 Truth Table | | |
|---|---|---|
| **B** | **A** | **Y** |
| 0 | 0 | |
| 0 | 1 | |
| 1 | 0 | |
| 1 | 1 | |

**Table 2-4**   7432 Quad Two-Input OR Truth Table

5. Open circuits file **d02-02c**. This circuit uses two sections of the 7432 IC to solve its logic problem (with three inputs). The Boolean equation for

   this circuit is _____. Actuate input switches A, B, and C and fill out Table 2-5 on page 16.

6. Open circuits file **d02-02d**. With this OR circuit you are going to view a square wave waveform using a virtual MultiSIM oscilloscope. The circuit is designed to inhibit the square wave when switch A is placed in the high position (5 V). When the switch is in the low position (GND), the square wave passes through the OR gate to the output terminal. To view this waveform, double-click on the oscilloscope and it will enlarge. Left-click on the workspace (don't miss this step or the circuit won't activate) and then toggle switch A. Observe the output of the OR gate displayed on the oscilloscope. When is the square wave signal inhibited? You should see that the square wave is inhibited when switch A is in the

   _____ position. The output of the OR gate is _____ V when the square wave is inhibited. The term **inhibited** refers to the trace on the

| 7432 Truth Table | | | |
|:---:|:---:|:---:|:---:|
| C | B | A | Y |
| 0 | 0 | 0 | |
| 0 | 0 | 1 | |
| 0 | 1 | 0 | |
| 0 | 1 | 1 | |
| 1 | 0 | 0 | |
| 1 | 0 | 1 | |
| 1 | 1 | 0 | |
| 1 | 1 | 1 | |

**Table 2-5**   Three-Input OR Gate Truth Table

oscilloscope being a constant 0 V or 5 V (a DC signal) and not changing with the square wave input.

## ● *Troubleshooting Problems:*

7. Open circuits file **d02-02e**. This circuit has a problem; there is no output according to the voltage probe. Use the DMM to locate the problem? The

   problem is _____

   _____.

8. Open circuits file **d02-02f**. This circuit has a problem; there is no output according to the voltage probe. Use the DMM to locate the problem? The

   problem is _____

   _____.

## ● *Circuit Construction:*

9. Open circuits file **d02-02g**. In this circuits file, you are to use the components on the workspace and connect the 7432 according to the Boolean equation $A + B + C + D + E = Y$ and Figure 2-5 (the circuit is partially connected). How many sections of the IC are left after you have

   completed this project? There is/are _____ section(s) left. Actuate the input switches A, B, C, D, and E, and enter the results into Table 2-6.

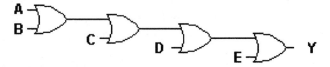

**Figure 2-5**   Five-Input OR Circuit Logic

| 7432 Truth Table | | | | | |
|---|---|---|---|---|---|
| **E** | **D** | **C** | **B** | **A** | **Y** |
| 0 | 0 | 0 | 0 | 0 | |
| 0 | 0 | 0 | 0 | 1 | |
| 0 | 0 | 0 | 1 | 0 | |
| 0 | 0 | 0 | 1 | 1 | |
| 0 | 0 | 1 | 0 | 0 | |
| 0 | 0 | 1 | 0 | 1 | |
| 0 | 0 | 1 | 1 | 0 | |
| 0 | 0 | 1 | 1 | 1 | |
| 0 | 1 | 0 | 0 | 0 | |
| 0 | 1 | 0 | 0 | 1 | |
| 0 | 1 | 0 | 1 | 0 | |
| 0 | 1 | 0 | 1 | 1 | |
| 0 | 1 | 1 | 0 | 0 | |
| 0 | 1 | 1 | 0 | 1 | |
| 0 | 1 | 1 | 1 | 0 | |
| 0 | 1 | 1 | 1 | 1 | |
| 1 | 0 | 0 | 0 | 0 | |
| 1 | 0 | 0 | 0 | 1 | |
| 1 | 0 | 0 | 1 | 0 | |
| 1 | 0 | 0 | 1 | 1 | |
| 1 | 0 | 1 | 0 | 0 | |
| 1 | 0 | 1 | 0 | 1 | |
| 1 | 0 | 1 | 1 | 0 | |
| 1 | 0 | 1 | 1 | 1 | |
| 1 | 1 | 0 | 0 | 0 | |
| 1 | 1 | 0 | 0 | 1 | |
| 1 | 1 | 0 | 1 | 0 | |
| 1 | 1 | 0 | 1 | 1 | |
| 1 | 1 | 1 | 0 | 0 | |
| 1 | 1 | 1 | 0 | 1 | |
| 1 | 1 | 1 | 1 | 0 | |
| 1 | 1 | 1 | 1 | 1 | |

**Table 2-6**   Five-Input OR Gate Truth Table

## Activity 2.3: The AND Gate

1. The **AND** gate is a circuit that provides a binary one or a HIGH at its output if all of its inputs are HIGH. If any of its inputs are binary zeros or LOW, then the output will be LOW. Figure 2-6 displays the logic symbol and truth table for all possible input and output conditions for a two-input AND gate. The Boolean expression for this AND gate is $A \cdot B = Y$. The · symbol indicates the AND function. Many times the equation will leave the · symbol out and the equation becomes $AB = Y$ (similar to the multiplication symbol used in standard algebra).

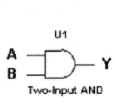

| Two-Input AND Gate Truth Table | | |
|:---:|:---:|:---:|
| B | A | Y |
| 0 | 0 | 0 |
| 0 | 1 | 0 |
| 1 | 0 | 0 |
| 1 | 1 | 1 |

**Figure 2-6**   Two-Input AND Gate Truth Table

2. Open circuits file **d02-03a**. Actuate the A input and B input switches according to the truth table and enter the results (Y) in Table 2-7. The voltage probe (X1) indicates the HIGH or LOW condition of output Y. The

Boolean equation for this circuit is _____

_____.

| AND Gate Truth Table | | |
|:---:|:---:|:---:|
| B | A | Y |
| 0 | 0 | |
| 0 | 1 | |
| 1 | 0 | |
| 1 | 1 | |

**Table 2-7**   AND Gate Truth Table

3. The pinout of a typical AND gate is shown in Figure 2-7. This AND gate is a 7408 Quad two-input IC consisting of four individual AND gates with each of the gates having two inputs. Open circuits file **d02-03b**. In this circuit, only the first AND gate is being used with a voltage probe connected to output 1Y to indicate the output condition of the gate. Actuate the input switches, A and B, and enter the results in Table 2-8.

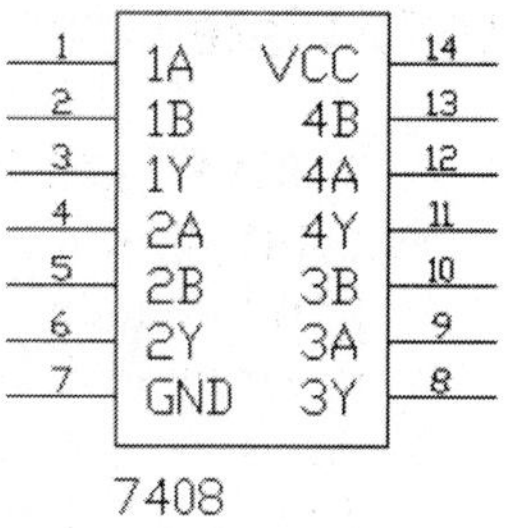

**Figure 2-7**   Pinout for a 7408 AND Gate

| 7408 Truth Table | | |
|:---:|:---:|:---:|
| **B** | **A** | **Y** |
| 0 | 0 | |
| 0 | 1 | |
| 1 | 0 | |
| 1 | 1 | |

**Table 2-8**   7408 Quad Two-Input AND Truth Table

4. Open circuits file **d02-03c**. This circuit uses two sections of the 7408 IC to solve its logic problem. The Boolean equation for this circuit is

_____. Actuate the input switches A, B, and C, and enter the resultant data in Table 2-9.

| 7408 Truth Table | | | |
|:---:|:---:|:---:|:---:|
| **C** | **B** | **A** | **Y** |
| 0 | 0 | 0 | |
| 0 | 0 | 1 | |
| 0 | 1 | 0 | |
| 0 | 1 | 1 | |
| 1 | 0 | 0 | |
| 1 | 0 | 1 | |
| 1 | 1 | 0 | |
| 1 | 1 | 1 | |

**Table 2-9**   Three-Input AND Gate Truth Table

5. Open circuits file **d02-03d**. In this circuits file, connect the 7408 according to the Boolean equation $A \cdot B \cdot C \cdot D \cdot E = Y$ and Figure 2-8. How many sections of the IC are left after you have completed this project?

There are _____ sections left. Actuate the input switches A, B, C, D, and E, and enter the results into Table 2-10.

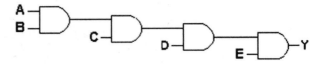

**Figure 2-8**    A Five-Input AND Circuit

| 7408 Truth Table | | | | | |
|:---:|:---:|:---:|:---:|:---:|:---:|
| **E** | **D** | **C** | **B** | **A** | **Y** |
| 0 | 0 | 0 | 0 | 0 | |
| 0 | 0 | 0 | 0 | 1 | |
| 0 | 0 | 0 | 1 | 0 | |
| 0 | 0 | 0 | 1 | 1 | |
| 0 | 0 | 1 | 0 | 0 | |
| 0 | 0 | 1 | 0 | 1 | |
| 0 | 0 | 1 | 1 | 0 | |
| 0 | 0 | 1 | 1 | 1 | |
| 0 | 1 | 0 | 0 | 0 | |
| 0 | 1 | 0 | 0 | 1 | |
| 0 | 1 | 0 | 1 | 0 | |
| 0 | 1 | 0 | 1 | 1 | |
| 0 | 1 | 1 | 0 | 0 | |
| 0 | 1 | 1 | 0 | 1 | |
| 0 | 1 | 1 | 1 | 0 | |
| 0 | 1 | 1 | 1 | 1 | |
| 1 | 0 | 0 | 0 | 0 | |
| 1 | 0 | 0 | 0 | 1 | |
| 1 | 0 | 0 | 1 | 0 | |
| 1 | 0 | 0 | 1 | 1 | |
| 1 | 0 | 1 | 0 | 0 | |
| 1 | 0 | 1 | 0 | 1 | |
| 1 | 0 | 1 | 1 | 0 | |

**Table 2-10**    Five-Input AND Gate Truth Table

| 7408 Truth Table | | | | | |
|---|---|---|---|---|---|
| **E** | **D** | **C** | **B** | **A** | **Y** |
| 1 | 0 | 1 | 1 | 1 | |
| 1 | 1 | 0 | 0 | 0 | |
| 1 | 1 | 0 | 0 | 1 | |
| 1 | 1 | 0 | 1 | 0 | |
| 1 | 1 | 0 | 1 | 1 | |
| 1 | 1 | 1 | 0 | 0 | |
| 1 | 1 | 1 | 0 | 1 | |
| 1 | 1 | 1 | 1 | 0 | |
| 1 | 1 | 1 | 1 | 1 | |

**Table 2-10**   (*continued*)

6. Open circuits file **d02-03e**. This circuit is designed to inhibit a digital signal when the switch is actuated. Actuate the switch and observe the output of the AND gate on the oscilloscope. When is the digital signal inhibited? The digital signal is inhibited when the switch is in the _____ V position. The output of the AND gate is _____ V when the digital signal is inhibited. The output of the gate is a _____ KHz signal when the gate is enabled. The gate is enabled when switch (A) is in the _____ V position.

## ● *Troubleshooting Problems:*

7. Open circuits file **d02-03f**. This circuit has a problem with switches A and B; the output voltage probe does not respond to these switches. Use the DMM or a voltage probe to troubleshoot the problem. The problem is

_____

_____.

8. Open circuits file **d02-03g.** This circuit has a problem; the voltage probe indicates a high output condition without switch C being actuated. What is the problem? The problem is _____

_____

_____.

## ● *Circuit Construction:*

9. Open circuits file **d02-03h**. Using the components and circuit on the workspace, construct a circuit that will illuminate lamp X1 when switches A and B are HIGH and switch C is LOW.

## Activity 2.4: The NAND Gate

1. The **NAND** gate is a circuit that provides a binary zero or a LOW at its output if all of its inputs are HIGH. If any of its inputs are binary zeros or LOW, then the output will be HIGH. The term NAND is a contraction of NOT and AND. Figure 2-9 shows the logic symbol for a two-input NAND gate and also a truth table for all possible input and output conditions. The Boolean expression for this NAND gate is $\overline{A \cdot B} = Y$. The logic symbol for the NAND is the same as the AND except for the bubble on the Y terminal representing inversion. The output is the complement (opposite) of an AND.

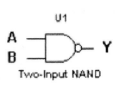

| NAND Truth Table | | |
|---|---|---|
| **B** | **A** | **Y** |
| 0 | 0 | 1 |
| 0 | 1 | 1 |
| 1 | 0 | 1 |
| 1 | 1 | 0 |

**Figure 2-9**   Two-Input NAND Gate Symbol and Truth Table

2. Open circuits file **d02-04a**. Actuate the A input and B input switches according to the truth table and enter the results (Y) in Table 2-11. The voltage probe indicates the HIGH or LOW condition of output Y. The

Boolean equation for this circuit is _____

_____.

| NAND Truth Table | | |
|---|---|---|
| **B** | **A** | **Y** |
| 0 | 0 | |
| 0 | 1 | |
| 1 | 0 | |
| 1 | 1 | |

**Table 2-11**   NAND Gate Truth Table

3. The pinout of a typical 7400 series NAND gate is shown in Figure 2-10. This NAND gate is a 7400 Quad two-input IC. It consists of four individual NAND gates with each gate having two inputs.

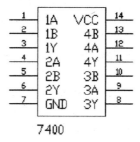

7400

**Figure 2-10**  Pinout for a 7400 NAND Gate

4. Open circuits file **d02-04b**. In this circuit, only the first NAND gate section of the IC is being used. It has a voltage probe connected to output 1Y to indicate the output logic condition of the gate. Actuate input switches A and B, and enter your data in Table 2-12. What is the Boolean equation for this gate? The Boolean equation for this

circuit is _____.

| 7400 Truth Table | | |
|---|---|---|
| **B** | **A** | **Y** |
| 0 | 0 | |
| 0 | 1 | |
| 1 | 0 | |
| 1 | 1 | |

**Table 2-12**  7400 Quad Two-Input NAND Truth Table

5. When trying to use a quad two-input NAND gate IC such as the 7400 for more than two inputs, the NAND logic of the first gate has to be inverted back to its previous logic level and then the result combined with the third input for the final NAND function. The inversion process usually takes place by using a two-input NAND gate as an inverter as displayed in Figure 2-11.

**U1A**

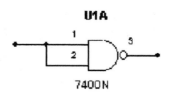

7400N

**Figure 2-11**  NAND Gate Inverter Circuit

6. Open circuits file **d02-04c**. This circuit demonstrates the use of a NAND gate as an inverter. Actuate the circuit and record the resultant data in Table 2-13.

| Switch Input | Output |
|---|---|
| **LOW (0 V)** | |
| **HIGH (5 V)** | |

**Table 2-13**   Using the NAND as an Inverter

7. Open circuits file **d02-04d**. This circuit demonstrates the procedure to be followed when there are more than two inputs and you are using two-input NAND ICs. The Boolean equation for this circuit is

_____. Actuate input switches A, B, and C and enter your results into Table 2-14.

| Three-Input NAND Truth Table | | | |
|---|---|---|---|
| **C** | **B** | **A** | **Y** |
| 0 | 0 | 0 | |
| 0 | 0 | 1 | |
| 0 | 1 | 0 | |
| 0 | 1 | 1 | |
| 1 | 0 | 0 | |
| 1 | 0 | 1 | |
| 1 | 1 | 0 | |
| 1 | 1 | 1 | |

**Table 2-14**   Three-Input NAND Gate Truth Table

8. Open circuits file **d02-04e**. This circuit uses one section of a dual four-input NAND IC, the 74LS20N, to monitor four inputs. Actuate input switches A, B, C, and D and enter the results into Table 2-15.

| 7420 Four-Input Truth Table | | | | |
|---|---|---|---|---|
| **D** | **C** | **B** | **A** | **Y** |
| 0 | 0 | 0 | 0 | |
| 0 | 0 | 0 | 1 | |
| 0 | 0 | 1 | 0 | |
| 0 | 0 | 1 | 1 | |
| 0 | 1 | 0 | 0 | |
| 0 | 1 | 0 | 1 | |
| 0 | 1 | 1 | 0 | |
| 0 | 1 | 1 | 1 | |
| 1 | 0 | 0 | 0 | |
| 1 | 0 | 0 | 1 | |
| 1 | 0 | 1 | 0 | |
| 1 | 0 | 1 | 1 | |
| 1 | 1 | 0 | 0 | |
| 1 | 1 | 0 | 1 | |
| 1 | 1 | 1 | 0 | |
| 1 | 1 | 1 | 1 | |

**Table 2-15**   Four-Input 7420 NAND Gate Truth Table

9. Two-input NAND gates cannot be directly used for four inputs as is done with two-input AND gates. The inversion process creates a problem using NAND gates in a manner similar to AND gates. It is possible to use AND gates in the place of two-input NAND gates and then invert the final output. It takes three two-input AND gates and a final inverter to replace one section of the 7420 IC. Figure 2-12 displays this type of circuit. How

many two-input AND ICs are needed for this circuit? _____ two-input AND gates are necessary.

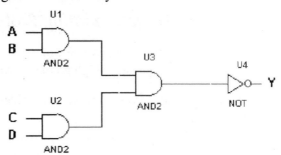

**Figure 2-12**   NAND Type Logic Diagram with
Four Inputs (using ANDs)

10. Open circuits file **d02-04f**. This circuit uses two-input NAND gates to replace a 7420 IC. How many two-input NAND gates are necessary to replace the 7420 IC? It is necessary to use _____ two-input NAND gates to replace the 7420 NAND IC. Activate the circuit and enter the resultant data in Table 2-16.

| 7420 Four-Input Truth Table | | | | |
|---|---|---|---|---|
| D | C | B | A | Y |
| 0 | 0 | 0 | 0 | |
| 0 | 0 | 0 | 1 | |
| 0 | 0 | 1 | 0 | |
| 0 | 0 | 1 | 1 | |
| 0 | 1 | 0 | 0 | |
| 0 | 1 | 0 | 1 | |
| 0 | 1 | 1 | 0 | |
| 0 | 1 | 1 | 1 | |
| 1 | 0 | 0 | 0 | |
| 1 | 0 | 0 | 1 | |
| 1 | 0 | 1 | 0 | |
| 1 | 0 | 1 | 1 | |
| 1 | 1 | 0 | 0 | |
| 1 | 1 | 0 | 1 | |
| 1 | 1 | 1 | 0 | |
| 1 | 1 | 1 | 1 | |

**Table 2-16**   Four-Input NAND Circuit Using Two-Input NAND

## ● *Troubleshooting Problems:*

11. Open circuits file **d02-04g**. This circuit has a connection problem; the output stays high no matter what the inputs do. Use the DMM or the voltage probe to solve the problem. The problem is _____

_____

_____.

12. Open circuits file **d02-04h**. This circuit has a problem; the output is not correct according to the voltage probe and expected truth table results.

    What is the problem? The problem is _____

    _____

    _____.

## ● *Circuit Construction:*

13. Open circuits file **d02-04i**. Using the logic diagram of Figure 2-12, construct a similar circuit on the workspace. Actuate logic switches A, B, C, and D and enter the results in Table 2-17. This project is typical of tasks that digital technicians are required to perform. Many times the technician is given a logic diagram of a circuit to build and simulate using the actual components. At that point, you go to the parts bin, get out the parts, and try to make the logic diagram function as it is intended to function.

| Four-Input Truth Table (NAND Circuit using ANDs) | | | | |
|---|---|---|---|---|
| D | C | B | A | Y |
| 0 | 0 | 0 | 0 | |
| 0 | 0 | 0 | 1 | |
| 0 | 0 | 1 | 0 | |
| 0 | 0 | 1 | 1 | |
| 0 | 1 | 0 | 0 | |
| 0 | 1 | 0 | 1 | |
| 0 | 1 | 1 | 0 | |
| 0 | 1 | 1 | 1 | |
| 1 | 0 | 0 | 0 | |
| 1 | 0 | 0 | 1 | |
| 1 | 0 | 1 | 0 | |
| 1 | 0 | 1 | 1 | |
| 1 | 1 | 0 | 0 | |
| 1 | 1 | 0 | 1 | |
| 1 | 1 | 1 | 0 | |
| 1 | 1 | 1 | 1 | |

**Table 2-17**  Truth Table for NAND Type Logic Diagram with Four Inputs

14. Open circuits file **d02-04j**. In this circuits file, you are to connect the 7420 IC (a dual four-input NAND IC) according to the Boolean equation $\overline{A \cdot B \cdot C \cdot D} = Y$. How many sections of the IC are left after you have completed this project? There are/is _____ section(s) left. Actuate input switches A, B, C, and D and enter the results into Table 2-18. Notice that each section of this IC has four inputs and no inverter is required.

| 7420 Four-Input Truth Table | | | | |
|---|---|---|---|---|
| **D** | **C** | **B** | **A** | **Y** |
| 0 | 0 | 0 | 0 | |
| 0 | 0 | 0 | 1 | |
| 0 | 0 | 1 | 0 | |
| 0 | 0 | 1 | 1 | |
| 0 | 1 | 0 | 0 | |
| 0 | 1 | 0 | 1 | |
| 0 | 1 | 1 | 0 | |
| 0 | 1 | 1 | 1 | |
| 1 | 0 | 0 | 0 | |
| 1 | 0 | 0 | 1 | |
| 1 | 0 | 1 | 0 | |
| 1 | 0 | 1 | 1 | |
| 1 | 1 | 0 | 0 | |
| 1 | 1 | 0 | 1 | |
| 1 | 1 | 1 | 0 | |
| 1 | 1 | 1 | 1 | |

**Table 2-18** Four-Input 7420 NAND Gate Truth Table

## Activity 2.5: The NOR Gate

1. The **NOR** gate is a circuit that provides a binary one or a HIGH at its output if all of its inputs are LOW. If any of its inputs are binary ones or HIGH, then the output will be LOW. The term NOR is a contraction of NOT and OR. Figure 2-13 displays the logic symbol for a two-input NOR gate and also a truth table for all possible input and output conditions. The Boolean algebra expression for this NOR gate is $\overline{A + B} = Y$. Notice that the symbol of the NOR gate is the same as the OR gate except for the bubble on the Y terminal representing inversion. The NOR gate output is the complement of an OR gate output for the same inputs.

| Truth Table for Two-Input NOR Gate | | |
|---|---|---|
| **B** | **A** | **Y** |
| 0 | 0 | 1 |
| 0 | 1 | 0 |
| 1 | 0 | 0 |
| 1 | 1 | 0 |

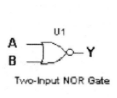

**Figure 2-13**  Two-Input NOR Gate Symbol and Truth Table

2. Open circuits file **d02-05a**. Actuate input switches A and B according to the truth table and place the results (Y) in Table 2-19. The voltage probe indicates the HIGH or LOW condition of output Y. The Boolean equation

for this circuit is _____.

| NOR Truth Table | | |
|---|---|---|
| **B** | **A** | **Y** |
| 0 | 0 | |
| 0 | 1 | |
| 1 | 0 | |
| 1 | 1 | |

**Table 2-19**  NOR Gate Truth Table

3. The pinout of a typical 7428 Quad two-input NOR gate is shown in Figure 2-14. It contains four individual NOR gates. Each individual gate has two inputs.

```
1   1Y    VCC  14
2   1A    4Y   13
3   1B    4B   12
4   2Y    4A   11
5   2A    3Y   10
6   2B    3B    9
7   GND   3A    8
        7428
```

**Figure 2-14**  Pinout for a 7428 NOR Gate

4. Open circuits file **d02-05b**. In this circuit, only the first section of the NOR gate is being used. The voltage probe connected to output 1Y indicates the output condition of the gate. Actuate the input switches A and B and enter your results in Table 2-20.

| 7428 Truth Table | | |
|---|---|---|
| B | A | Y |
| 0 | 0 | |
| 0 | 1 | |
| 1 | 0 | |
| 1 | 1 | |

**Table 2-20** 7428 Quad Two-Input NOR Truth Table

5. Open circuits file **d02-05c**. This three-input circuit uses three sections of the 7428 IC to solve its logic problem with one section being used as an inverter. The Boolean equation for this three-input NOR circuit is

_____. Actuate input switches A, B, and C and enter your results in Table 2-21.

| Three-Input NOR Circuit Truth Table | | | |
|---|---|---|---|
| C | B | A | Y |
| 0 | 0 | 0 | |
| 0 | 0 | 1 | |
| 0 | 1 | 0 | |
| 0 | 1 | 1 | |
| 1 | 0 | 0 | |
| 1 | 0 | 1 | |
| 1 | 1 | 0 | |
| 1 | 1 | 1 | |

**Table 2-21** Three-Input NOR Circuit Truth Table

6. In a manner similar to the NAND gate, the NOR gate can be used as an inverter (NOT gate) by connecting its inputs together. Also, in a manner similar to the NAND gate, the logic of the first two-input gate has to be inverted before the third input (switch C) can become part of the solution to the NOR logic problem.

7. Open circuits file **d02-05d**. In this circuit, 74LS28N Quad two-input ICs are being used for four inputs. Actuate the switches and enter the resultant data in Table 2-22. Two NOR ICs had to be used for this circuit. How

   many sections are left over in the second IC? There are _____ sections left over in the second IC.

| D | C | B | A | Y |
|---|---|---|---|---|
| 0 | 0 | 0 | 0 |   |
| 0 | 0 | 0 | 1 |   |
| 0 | 0 | 1 | 0 |   |
| 0 | 0 | 1 | 1 |   |
| 0 | 1 | 0 | 0 |   |
| 0 | 1 | 0 | 1 |   |
| 0 | 1 | 1 | 0 |   |
| 0 | 1 | 1 | 1 |   |
| 1 | 0 | 0 | 0 |   |
| 1 | 0 | 0 | 1 |   |
| 1 | 0 | 1 | 0 |   |
| 1 | 0 | 1 | 1 |   |
| 1 | 1 | 0 | 0 |   |
| 1 | 1 | 0 | 1 |   |
| 1 | 1 | 1 | 0 |   |
| 1 | 1 | 1 | 1 |   |

**Table 2-22**   Four Input Circuit Using Quad Two-Input NOR ICs

8. Open circuits file **d02-05e**. In this file, you have two 7428 ICs connected according to the Boolean equation $\overline{A + B + C + D} = Y$ and following the logic of Figure 2-15. How many sections of the ICs are left after you have

   completed this project? There are _____ sections left. Actuate the input switches A, B, C, and D and enter the results into Table 2-23.

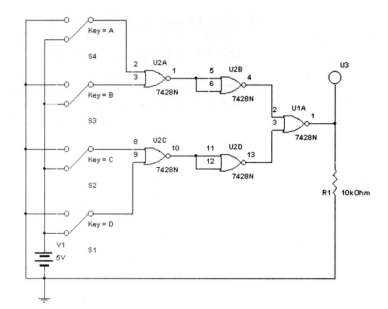

**Figure 2-15**   A Four-Input NOR Circuit

| D | C | B | A | Y |
|---|---|---|---|---|
| 0 | 0 | 0 | 0 | |
| 0 | 0 | 0 | 1 | |
| 0 | 0 | 1 | 0 | |
| 0 | 0 | 1 | 1 | |
| 0 | 1 | 0 | 0 | |
| 0 | 1 | 0 | 1 | |
| 0 | 1 | 1 | 0 | |
| 0 | 1 | 1 | 1 | |
| 1 | 0 | 0 | 0 | |
| 1 | 0 | 0 | 1 | |
| 1 | 0 | 1 | 0 | |
| 1 | 0 | 1 | 1 | |
| 1 | 1 | 0 | 0 | |
| 1 | 1 | 0 | 1 | |
| 1 | 1 | 1 | 0 | |
| 1 | 1 | 1 | 1 | |

**Table 2-23**   Four-Input NOR Circuit Truth Table

## ● *Troubleshooting Problems:*

9. Open circuits file **d02-05f**. This circuit is supposed to be a normal two-input NOR circuit and should operate according to the truth table of Figure 2-16. Check the operation of the circuit according to the truth table. Is there a problem and what is it? The problem is _____

   _____

   _____.

| NOR Truth Table | | |
|---|---|---|
| **B** | **A** | **Y** |
| 0 | 0 | 1 |
| 0 | 1 | 0 |
| 1 | 0 | 0 |
| 1 | 1 | 0 |

**Figure 2-16**   Two-Input NOR Gate
Truth Table

10. Open circuits file **d02-05g**. This circuit has a component problem; the C and D switches seem to have no effect on the output. What is the problem? The problem is _____

    _____

    _____.

## ● *Circuit Construction:*

11. Open circuits file **d02-05h**. In this circuit file, you are to connect the circuit according to the Boolean equation $\overline{A + B} = Y$. Activate your circuit and enter the resultant data in Table 2-24 on page 34. Does the truth table agree with Figure 2-16 (it should)?

**NOR Truth Table**

| B | A | Y |
|---|---|---|
| 0 | 0 | |
| 0 | 1 | |
| 1 | 0 | |
| 1 | 1 | |

**Table 2-24**   Constructed Two-Input NOR Circuit Data

12. Open circuits file **d02-05i**. In this circuit file, you are to connect the circuit according to the Boolean equation $\overline{A+B+C} = Y$. Activate your circuit and enter the resultant data in Table 2-25. Does the truth table agree with your previous three-input NOR circuits (it should)?

**Three-Input NOR Circuit Truth Table**

| D | B | A | Y |
|---|---|---|---|
| 0 | 0 | 0 | |
| 0 | 0 | 1 | |
| 0 | 1 | 0 | |
| 0 | 1 | 1 | |
| 1 | 0 | 0 | |
| 1 | 0 | 1 | |
| 1 | 1 | 0 | |
| 1 | 1 | 1 | |

**Table 2-25**   Three-Input NOR Circuit Truth Table

## Activity 2.6: The Buffer Gate

1. The **Buffer** gate is used to change the electrical characteristics of a digital circuit without changing its logic level characteristics. The buffer gate provides isolation between components, physical circuit locations, and other logic circuits. The buffer also provides current amplification to expand the output current capabilities of other logic gates with the low current input becoming a higher current output. Figure 2-17 displays the logic symbol and truth table for a buffer.

| Buffer Gate Truth Table | |
|---|---|
| **A** | **Y** |
| 0 | |
| 1 | |

**Figure 2-17**   B uffer Gate Symbol and Truth Table

2. Open circuits file **d02-06a**. This circuits file displays various buffer and inverter options available to the digital circuit designer. U___, U___, and U___ are the inverter (NOT) gates, and U___, U___, and U___ are the buffer gates. Actuate the input switch and record the resultant logic data in Table 2-26.

| Various Buffer and Inverter Gate Truth Table | | | | | | |
|---|---|---|---|---|---|---|
| **A**<br>**Input** | **U1**<br>**NOT**<br>**Output** | **U3**<br>**NAND**<br>**Output** | **U5**<br>**NOR**<br>**Output** | **U7**<br>**AND**<br>**Output** | **U9**<br>**OR**<br>**Output** | **U11**<br>**Buffer**<br>**Output** |
| 0 | | | | | | |
| 1 | | | | | | |

**Table 2-26**   Truth Table for Various Buffer and Inverter Gates

3. Open circuits file **d02-06b**. Actuate switch A and enter the resultant data for the buffer circuit in Table 2-27.

| Buffer Gate Truth Table | |
|---|---|
| **A** | **Y** |
| 0 | |
| 1 | |

**Table 2-27**   Buffer Gate Truth Table

# 3. Advanced Logic Gates

**References**

*MultiSIM* 2001

*MultiSIM* 2001 Study Guide

**Objectives**   After completing this chapter, you should be able to:

- Use MultiSIM® circuits to perform advanced logic operations.
- Determine truth table information for advanced logic gates.
- Use NAND and NOR gates as universal gates.
- Use Boolean algebra to construct advanced logic circuits.
- Use Exclusive gates to implement Boolean algebra equations.
- Construct logic circuits.
- Troubleshoot logic circuits.

## Introduction

### Advanced Logic Circuits

The designation, advanced logic circuits, refers to digital logic circuits that are more complex than the basic gates. These complex circuits are constructed as various combinations of basic logic circuits where specific tasks are obtained through designated connections. An example of these designated connections is the category of logic gates referred to as universal gates where NAND and NOR gates are used explicitly to replace any other logic gate. In addition to the universal gates, two of the advanced logic gates that we will study in this chapter are:

- The **Exclusive-OR (XOR)** gate

- The **Exclusive-NOR (XNOR)** gate

Inside the integrated circuits, these two exclusive gates consist of various combinations of the basic logic gates that are necessary to perform the required tasks.

## Activity 3.1: The Exclusive OR (XOR) Gate

1. The **XOR** logic gate is identical to the OR gate with one exception, whenever the inputs are the same (HIGHs or LOWs), the output will always be LOW. The logic symbol and the basic logic circuit for a two-input XOR gate is shown in Figure 3-1. Figure 3-2 displays the truth

table for the circuit. The logic circuit meets the requirements of the Boolean equation for the XOR gate. The NOT function for the circuit is symbolized by the ′ symbol in the MultiSIM diagram. The Boolean equation for this XOR gate would be $A \oplus B = Y$. The Boolean symbol for the XOR function is an encircled OR symbol, $\oplus$.

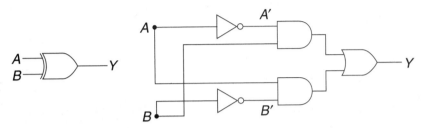

**Figure 3-1**   Two-Input XOR Logic Symbol and XOR Logic Circuit

| Two-Input XOR Truth Table for $A \oplus B = Y$ | | |
|:---:|:---:|:---:|
| **B** | **A** | **Y** |
| 0 | 0 | 0 |
| 0 | 1 | 1 |
| 1 | 0 | 1 |
| 1 | 1 | 0 |

**Figure 3-2**   Example of Two-Input XOR Truth Table

2.  Open circuits file **d03-01a**. Actuate switches A and B according to the truth table and fill the resultant data (Y) into Table 3-1. The voltage probe indicates the HIGH or LOW condition of output Y. The Boolean equation

for this circuit is _____.

| Two-Input XOR Truth Table | | |
|:---:|:---:|:---:|
| **B** | **A** | **Y** |
| 0 | 0 | |
| 0 | 1 | |
| 1 | 0 | |
| 1 | 1 | |

**Table 3-1**   Two-Input XOR Truth Table

3. The pinout of a 7400 series XOR gate is shown in Figure 3-3. This XOR gate is a 7486 Quad two-input IC consisting of four separate XOR gates with each gate having two inputs.

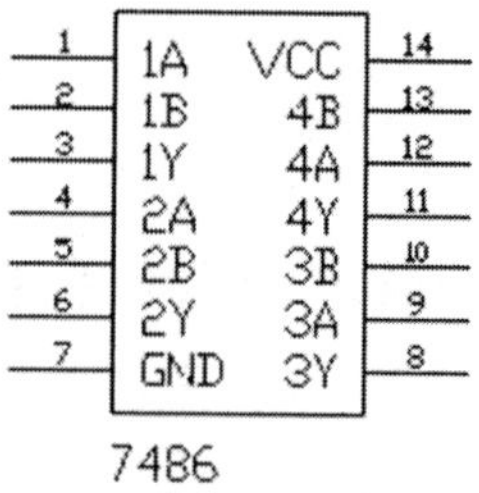

7486

**Figure 3-3** Pinout for a 7486 XOR Gate

4. Open circuits file **d03-01b**. In this circuit, only the first XOR gate (U1A) is being used. It has a voltage probe connected to the output of U1A to indicate the HIGH or LOW output condition of the gate. Actuate input switches A and B and enter your data into Table 3-2. The Boolean

equation for this circuit is _____. The 7486 IC can only be used in the two-input mode; the unused two-input gates cannot be used to expand XOR inputs. XOR gates with more than two inputs are necessary when more than two inputs are required. Other basic logic gates are normally used to provide XOR logic for more than two inputs.

| 7486 Truth Table | | |
|---|---|---|
| B | A | Y |
| 0 | 0 | |
| 0 | 1 | |
| 1 | 0 | |
| 1 | 1 | |

**Table 3-2** 7486 Quad Two-Input XOR Truth Table

## ● *Troubleshooting Problems:*

5. Open circuits file **d03-01c**. This circuit has a problem; the output does not change according to a typical two-input XOR truth table. What is the

problem? The problem is _____

_____

_____.

6. Open circuits file **d03-01d**. This circuit is not working correctly. The output doesn't change at all. What is the problem? The problem is _____

_____

_____.

● *Circuit Construction:*

7. Open circuits file **d03-01e** and use the components on the workspace to construct a two-input XOR circuit according to Figure 3-4. Activate the circuit and enter the resultant data in Table 3-3.

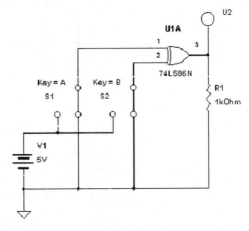

**Figure 3-4**   74LS86N Exclusive OR Circuit

| 7486 Truth Table | | |
|---|---|---|
| **B** | **A** | **Y** |
| 0 | 0 | |
| 0 | 1 | |
| 1 | 0 | |
| 1 | 1 | |

**Table 3-3**   Truth Table for Constructed
7486 Quad Two-Input Circuit

## Activity 3.2:  The Exclusive NOR (XNOR) Gate

1. The **XNOR** logic gate is identical to the NOR logic gate with one exception, whenever the inputs are the same (HIGHs or LOWs), the output will be HIGH. The XNOR is the complement of the XOR. The logic symbol and the basic logic circuit for a two-input XNOR gate is shown in Figure 3-5. Figure 3-6 displays the truth table for the circuit. Notice that the logic circuit meets the requirements of the Boolean equation for

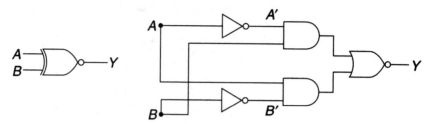

**Figure 3-5**   XNOR Logic Symbol and XNOR Logic Circuit

| Two-Input XNOR Truth Table | | |
|---|---|---|
| **B** | **A** | **Y** |
| 0 | 0 | 1 |
| 0 | 1 | 0 |
| 1 | 0 | 0 |
| 1 | 1 | 1 |

**Figure 3-6**   XNOR Truth Table

the XNOR gate and that the NOT function is symbolized by the
$'$ symbol. The Boolean equation for this XNOR gate would be
$\overline{A \oplus B} = Y$, the NOT function of an XOR gate.

2. Open circuits file **d03-02a**. Activate the switches according to the truth
table and enter the resultant data (Y) in Table 3-4. The voltage probe indicates the HIGH or LOW condition of output Y. The Boolean equation

for this circuit is _____.

| XNOR Truth Table | | |
|---|---|---|
| **B** | **A** | **Y** |
| 0 | 0 | |
| 0 | 1 | |
| 1 | 0 | |
| 1 | 1 | |

**Table 3-4**   XNOR Truth Table

3. One simple way to provide the necessary logic for an XNOR circuit is to
complement the output of an XOR circuit. This could be accomplished
by using an XOR circuit in series with an inverter (NOT) circuit.

4. Open circuits file **d03-02b**. In this circuit, the 74LS266N XNOR IC is
used to provide XNOR logic. Activate the switches according to the truth

table and place the resultant data (Y) in Table 3-5. The voltage probe indicates the HIGH or LOW condition of output Y. The Boolean equation for this circuit is _____. Notice the extra resistor, R2, connected to +5 V. This resistor provides the power for the 74LS266N IC; the 74LS266N IC is an open collector IC and needs external power to operate correctly. We will study open collector ICs in chapter 5.

| XNOR Truth Table | | |
|---|---|---|
| B | A | Y |
| 0 | 0 | |
| 0 | 1 | |
| 1 | 0 | |
| 1 | 1 | |

**Table 3-5**  XNOR Truth Table

## ● *Troubleshooting Problems:*

5. Open circuits file **d03-02c**. This circuit has a problem; the output does not change according to a typical two-input XNOR truth table. What is the problem? The problem is _____

_____

_____.

6. Open circuits file **d03-02d**. Another technician built this XNOR circuit, and it does not operate according to an XNOR truth table. Figure 3-7 displays the test data for the circuit. What is wrong with the circuit? The problem with the circuit is _____

_____

_____.

| XNOR Circuit Truth Table | | | |
|---|---|---|---|
| B | A | Predicted Data | Actual Data |
| 0 | 0 | 1 | 0 |
| 0 | 1 | 0 | 1 |
| 1 | 0 | 0 | 1 |
| 1 | 1 | 1 | 0 |

**Figure 3-7**  Data for Tested Circuit (XNOR)

● *Circuit Construction:*

7. Open circuits file **d03-02e** and construct a two-input XNOR circuit using the components on the workspace according to Figure 3-8. Activate the circuit and enter the resultant data in Table 3-6.

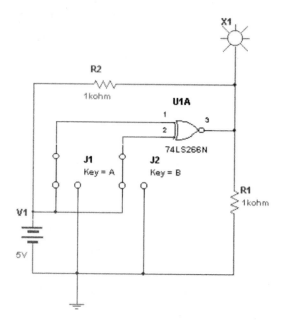

**Figure 3-8**  Two-Input XNOR Circuit

| XNOR Truth Table | | |
|---|---|---|
| B | A | Y |
| 0 | 0 | |
| 0 | 1 | |
| 1 | 0 | |
| 1 | 1 | |

**Table 3-6**  Truth Table for Constructed XNOR Circuit

## Activity 3.3: The NAND as a Universal Gate

1. The NAND logic gate (along with the NOR logic gate) can be used in various combinations to replace any other logic gate and has earned the title of **Universal Gate**.

2. You have already been introduced to the use of the NAND gate as an inverter. Open circuits file **d03-03a**. This circuit uses all of the gates in a NAND IC as inverters. Use the DMM to measure the voltages at the output of each gate and enter the data in Table 3-7.

| | Output Pin 3 | Output Pin 6 | Output Pin 8 | Output Pin 11 |
|---|---|---|---|---|
| **HIGH Input** | | | | |
| **LOW Input** | | | | |

**Table 3-7**   Using the NAND as an Inverter

3. It takes three two-input NAND gates to replace a two-input OR gate as displayed in Figure 3-9. Open circuits file **d03-03b**; compare the outputs of the two circuits, and enter your data in Table 3-8. The data entered in the OR column should be the same as the data entered in the NAND/OR column.

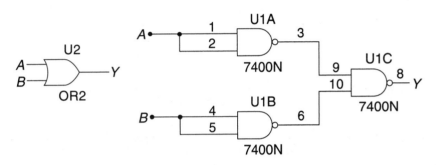

**Figure 3-9**   Replacing an OR Gate with NAND Gates

| OR Truth Table | | | |
|---|---|---|---|
| **B Input** | **A Input** | **OR Gate Output (Y1)** | **NAND/OR Gate Output (Y2)** |
| 0 | 0 | | |
| 0 | 1 | | |
| 1 | 0 | | |
| 1 | 1 | | |

**Table 3-8**   Comparing OR Circuit Configurations of Figure 3-9

4. It takes two two-input NAND gates to replace a two-input AND gate as shown in Figure 3-10. Open circuits file **d03-03c**; compare the outputs of the two circuits, and enter your data in Table 3-9. The data entered in the AND column should be the same as the data entered in the NAND/AND column.

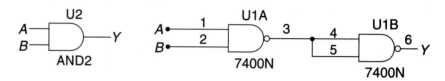

**Figure 3-10**   Replacing an AND Gate with NAND Gates

| AND Truth Table | | | |
|---|---|---|---|
| **B Input** | **A Input** | **AND Gate Output (Y1)** | **NAND/AND Gate Output (Y2)** |
| 0 | 0 | | |
| 0 | 1 | | |
| 1 | 0 | | |
| 1 | 1 | | |

**Table 3-9**   Comparing AND Circuit Configurations

5. It requires four two-input NAND gates to replace a two-input NOR gate as shown in Figure 3-11. Open circuits file **d03-03d**; compare the outputs of the two circuits, and enter your data in Table 3-10. The data entered in the NOR column should be the same as the data entered in the NAND/NOR column.

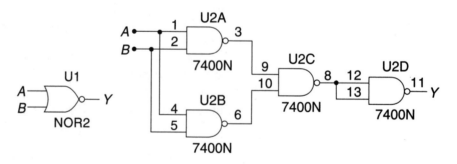

**Figure 3-11**   Replacing a NOR Gate with NAND Gates

| NOR Truth Table | | | |
|---|---|---|---|
| **B Input** | **A Input** | **NOR Gate Output (Y1)** | **NAND/NOR Gate Output (Y2)** |
| 0 | 0 | | |
| 0 | 1 | | |
| 1 | 0 | | |
| 1 | 1 | | |

**Table 3-10**   Comparing NOR Circuit Configurations

6. It takes four two-input NAND gates to replace a two-input XOR gate as shown in Figure 3-12. Open circuits file **d03-03e**; compare the outputs of the two circuits, and enter your data in Table 3-11. The data entered in the XOR column should be the same as the data entered in the NAND/XOR column.

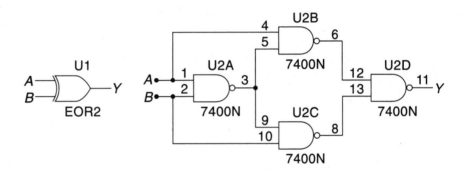

**Figure 3-12**   Replacing an XOR Gate with NAND Gates

| XOR Truth Table | | | |
|---|---|---|---|
| **B Input** | **A Input** | **XOR Gate Output (Y1)** | **NAND/XOR Gate Output (Y2)** |
| 0 | 0 | | |
| 0 | 1 | | |
| 1 | 0 | | |
| 1 | 1 | | |

**Table 3-11**   Comparing XOR Circuit Configurations

7. It takes five two-input NAND gates to replace a two-input XNOR gate as shown in Figure 3-13. Open circuits file **d03-03f** and compare the outputs of the two circuits. Enter the resultant data in Table 3-12. The data entered in the XNOR column should be the same as the data entered in the NAND/XNOR column.

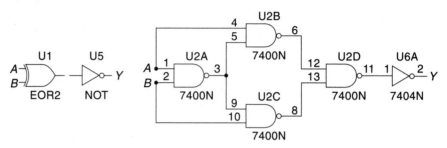

**Figure 3-13** Replacing an XNOR Gate with NAND Gates

| XNOR Truth Table | | | |
|---|---|---|---|
| **B Input** | **A Input** | **XNOR Gate Output (Y1)** | **NAND/XNOR Gate Output (Y2)** |
| 0 | 0 | | |
| 0 | 1 | | |
| 1 | 0 | | |
| 1 | 1 | | |

**Table 3-12** Comparing XNOR Circuit Configurations

8. When two NAND gates are connected as inverters (NOT gates) in series, they effectively cancel each other out (a NOT cancels out a NOT). That means that a circuit of this type can be simplified by removing any two inverters connected in series. As you can see in Figure 3-14, there are two pairs of NAND inverters in series (U2B-U2C and U1A-U1B). Remove them and the circuit should operate the same.

9.  Open circuits file **d03-03g**. This is the circuit of Figure 3-14. Activate the circuit and record the resultant data according to Table 3-13.

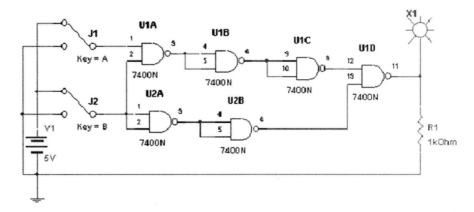

**Figure 3-14**   Logic Circuit According to $AB + \overline{B} = Y$

| Truth Table for $AB + \overline{B} = Y$ | | |
|---|---|---|
| **B Input** | **A Input** | **Output (Y)** |
| 0 | 0 | |
| 0 | 1 | |
| 1 | 0 | |
| 1 | 1 | |

**Table 3-13**   Truth Table for $AB + \overline{B} = Y$

10. Modify circuit **d03-03g**, removing inverters connected to inverters (for example, U1B connected to U1C). Activate the switches for the new circuit and enter the resultant data into Table 3-14. Compare the entries with Table 3-13; they should be the same. Draw your new circuit on Figure 3-15. Do not save the new circuit.

| Truth Table for $AB + \overline{B} = Y$ | | |
|---|---|---|
| **B Input** | **A Input** | **Output (Y)** |
| 0 | 0 | |
| 0 | 1 | |
| 1 | 0 | |
| 1 | 1 | |

**Table 3-14**   Truth Table for $AB + \overline{B} = Y$

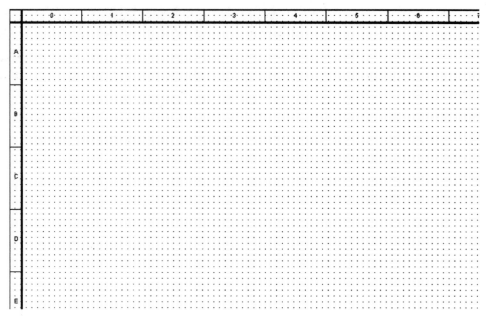

**Figure 3-15** Drawing of Revised Circuit According to $AB + \overline{B} = Y$

## ● *Troubleshooting Problems:*

11. Open circuits file **d03-03h**. In this circuit, a NAND/AND, the output is wrong. Actuate switches A and B and enter your results in Table 3-15. Use the DMM to troubleshoot the circuit. What is wrong?

The problem is _____

_____

_____.

| AND Truth Table | | |
|---|---|---|
| B Input | A Input | NAND/AND Gate Output (Y) |
| 0 | 0 | |
| 0 | 1 | |
| 1 | 0 | |
| 1 | 1 | |

**Table 3-15** NAND/AND Truth Table

12. Open circuits file **d03-03i**. In this circuit, a NAND/XNOR, the output is wrong. Actuate switches A and B and enter your results in Table 3-16. Use the DMM to troubleshoot the circuit. What is wrong?

| XNOR Truth Table | | |
|---|---|---|
| **B Input** | **A Input** | **NAND/XNOR Gate Output (Y)** |
| 0 | 0 | |
| 0 | 1 | |
| 1 | 0 | |
| 1 | 1 | |

**Table 3-16**   NAND/XNOR Truth Table

The problem is _____

_____

_____.

## ● *Circuit Construction:*

13. Open circuits file **d03-03j**. Using the universal techniques that you have learned with NAND gates, construct a combinational circuit according to the Boolean formula $A + \overline{AB} = Y$ from the components and ICs on the workspace. Operate the circuit and enter your data into Table 3-17. Your circuit should be similar to the circuit of Step 9 in this activity.

| Truth Table for $A + \overline{AB} = Y$ | | |
|---|---|---|
| **B Input** | **A Input** | **Output (Y)** |
| 0 | 0 | |
| 0 | 1 | |
| 1 | 0 | |
| 1 | 1 | |

**Table 3-17**   Truth Table for $A + \overline{AB} = Y$

## Activity 3.4: The NOR as a Universal Gate

1. The NOR logic gate can be used to replace any other logic gate, and along with the NAND gate, has also earned the title of Universal Gate. You have already been introduced to the use of the NOR gate as an inverter in a previous chapter.

2. Open circuits file **d03-04a**. This circuit uses all of the gates in a NOR IC as inverters. Measure the output voltage of each of the gates and enter the data in Table 3-18.

|  | Output Pin 1 | Output Pin 4 | Output Pin 10 | Output Pin 13 |
|---|---|---|---|---|
| **HIGH Input** |  |  |  |  |
| **LOW Input** |  |  |  |  |

**Table 3-18** Using the NOR as an Inverter

3. It requires two two-input NOR gates to replace a two-input OR gate as displayed in Figure 3-16. Open circuits file **d03-04b**; compare the outputs of the two circuits, and enter your data in Table 3-19. The data entered in the OR column should be the same as the data entered in the NOR/OR column.

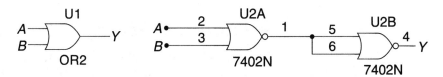

**Figure 3-16** Replacing an OR Gate with NOR Gates

| NOR/OR Truth Table | | | |
|---|---|---|---|
| **B Input** | **A Input** | **OR Gate Output (Y1)** | **NOR/OR Gate Output (Y2)** |
| 0 | 0 |  |  |
| 0 | 1 |  |  |
| 1 | 0 |  |  |
| 1 | 1 |  |  |

**Table 3-19** Comparing OR Circuit Configurations

4. It requires three two-input NOR gates to replace a two-input AND gate as shown in Figure 3-17. Open circuits file **d03-04c**; compare the outputs of the two circuits, and enter your data in Table 3-20. The data entered in the AND column should be the same as the data entered in the NOR/AND column.

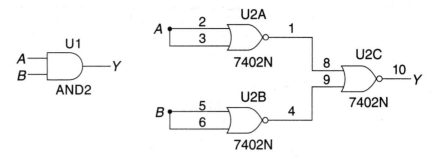

**Figure 3-17**   Replacing an AND Gate with NOR Gates

| NOR/AND Truth Table | | | |
|---|---|---|---|
| **B Input** | **A Input** | **AND Gate Output (Y1)** | **NOR/AND Gate Output (Y2)** |
| 0 | 0 | | |
| 0 | 1 | | |
| 1 | 0 | | |
| 1 | 1 | | |

**Table 3-20**   Comparing AND Circuit Configurations

5. It requires four two-input NOR gates to replace a two-input NAND gate as shown in Figure 3-18. Open circuits file **d03-04d**; compare the outputs of the two circuits, and enter your data in Table 3-21. The data entered in the NAND column should be the same as the data entered in the NOR/NAND column.

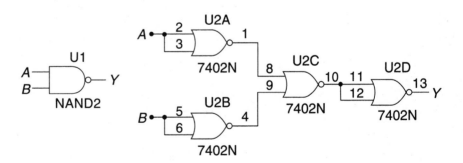

**Figure 3-18**   Replacing a NAND Gate with NOR Gates

| NOR/NAND Truth Table | | | |
|---|---|---|---|
| **B Input** | **A Input** | **NAND Gate Output (Y1)** | **NOR/NAND Gate Output (Y2)** |
| 0 | 0 | | |
| 0 | 1 | | |
| 1 | 0 | | |
| 1 | 1 | | |

**Table 3-21**    Comparing NOR/NAND Circuit Configurations

## ● *Troubleshooting Problems:*

6. Open circuits file **d03-04e**. This circuit has component problems. It appears to be constructed correctly, but the Y2 output is not working. What is the problem? The problem is _____

_____

_____.

7. Open circuits file **d03-04f**. This circuit has a connection problem; output X2 stays LOW when it should go HIGH. What is the problem? The problem is _____

_____

_____.

## ● *Circuit Construction:*

8. Open circuits file **d03-04g** and use NOR gates to construct a NAND circuit as displayed in Figure 3-18. Actuate switches A and B and enter the resultant data in Table 3-22.

| NOR/NAND Truth Table | | |
|:---:|:---:|:---:|
| **B Input** | **A Input** | **NOR/NAND Gate Output (Y)** |
| 0 | 0 | |
| 0 | 1 | |
| 1 | 0 | |
| 1 | 1 | |

**Table 3-22**   Truth Table Data for Student-Constructed Circuit

## Activity 3.5: Working with Boolean Formulas and Combinational Circuits

1. In this activity you will be constructing digital circuits from Boolean formulas and determining Boolean formulas from already constructed digital circuits.

2. Open circuits file **d03-05a**. Use the components to construct a circuit according to the Boolean formula $A + B + C = Y$. Draw the circuit on Figure 3-19.

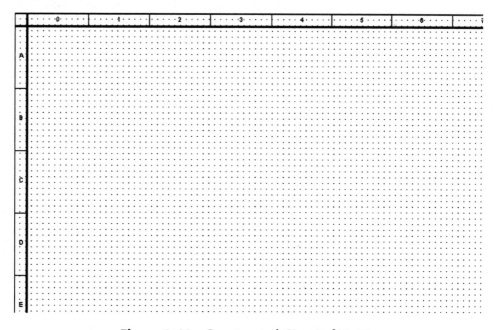

**Figure 3-19**   Constructed Circuit **d03-05a**

3. Open circuits file **d03-05b**. Use the components to construct a circuit according to the Boolean formula $A \cdot B \cdot C = Y$. Draw the circuit on Figure 3-20.

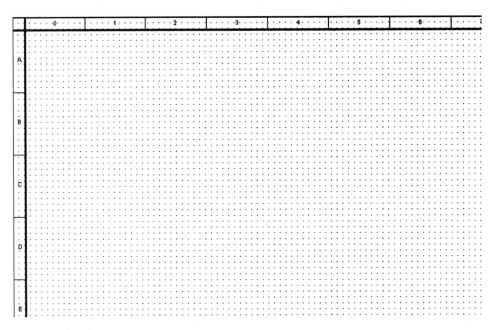

**Figure 3-20**   Constructed Circuit **d03-05b**

4. Open circuits file **d03-05c**. There are two circuits to be constructed in this step. Connect them in parallel according to the Boolean formulas $A (B + C) = Y$ and $A \cdot B + A \cdot C = Y$. Draw the two circuits on Figures 3-21 and 3-22.

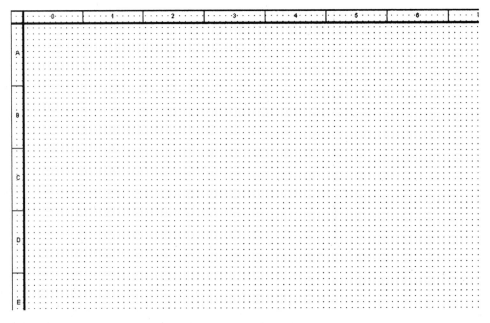

**Figure 3-21**   Constructed Circuit (1 of 2) **d03-05c** for: $A (B + C) = Y$

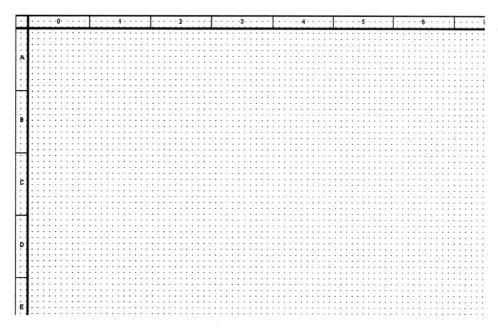

**Figure 3-22**   Constructed Circuit (2 of 2) **d03-05c** for: $A \cdot B + A \cdot C = Y$

5. Open circuits file **d03-05d**. Use the components to construct a circuit according to the Boolean formula $(A \cdot \overline{B} \cdot C) + (\overline{A} \cdot B \cdot \overline{C}) = Y$. Draw the circuit on Figure 3-23.

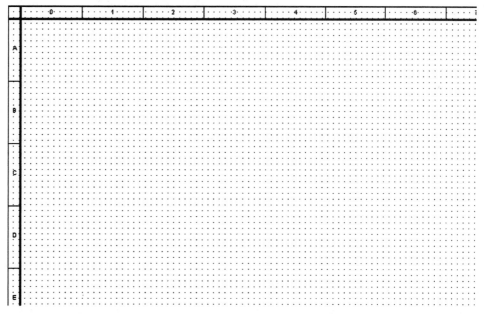

**Figure 3-23**   Constructed Circuit **d03-05d** for: $(A \cdot \overline{B} \cdot C) + (\overline{A} \cdot B \cdot \overline{C}) = Y$

6. Open circuits file **d03-05e**. Use the components to construct a circuit according to the Boolean formula $(A + B\overline{C})(\overline{AB} + C) = Y$. Draw the circuit on Figure 3-24. Then activate the circuit according to the truth table and enter the resultant data in Table 3-23.

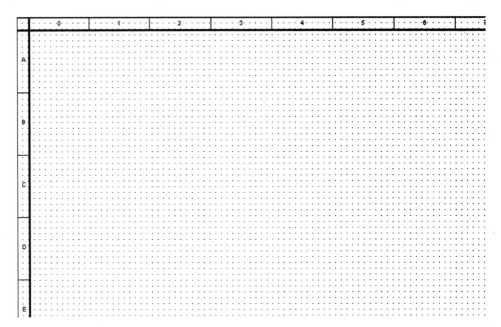

**Figure 3-24**   Constructed Circuit **d03-05e** for: $(A + B\overline{C})(\overline{AB} + C) = Y$

| Truth Table for Boolean Formula $(A + B\overline{C})(\overline{AB} + C) = Y$ | | | |
|---|---|---|---|
| **Input C** | **Input B** | **Input A** | **Output Y** |
| 0 | 0 | 0 | |
| 0 | 0 | 1 | |
| 0 | 1 | 0 | |
| 0 | 1 | 1 | |
| 1 | 0 | 0 | |
| 1 | 0 | 1 | |
| 1 | 1 | 0 | |
| 1 | 1 | 1 | |

**Table 3-23**   Truth Table for $(A + B\overline{C})(\overline{AB} + C) = Y$

7. Open circuits file **d03-05f**. Determine the Boolean equation for this circuit? The Boolean equation for this circuit is _____

_____.

8. Open circuits file **d03-05g**. Determine the Boolean equation for this circuit? The Boolean equation for this circuit is _____

_____.

# 4. Arithmetic Circuits

**References**

*MultiSIM* 2001

*MultiSIM* 2001 Study Guide

**Objectives**   After completing this chapter, you should be able to:

- Construct and use adder circuits.
- Troubleshoot adder circuits.
- Subtract with 1's and 2's complement circuits.
- Troubleshoot 1's and 2's complement circuits.

## Introduction

**Arithmetic** circuits, and specifically **Adder** circuits, are combinational circuits that perform mathematical functions including subtraction, multiplication, and division.

Some examples of adder ICs are:

- The 7483N 4-bit Binary Adder

  - Nine Inputs:  four A inputs (A1 through A4), four B inputs (B1 through B4) and a Carry input (C0)

  - Five Outputs:  four summation outputs (S1 through S4) and a Carry output (C4)

- The 4008BT 4-bit Full Adder

  - Nine Inputs:  four A inputs (A0 through A3), four B inputs (B0 through B3), and a Carry input (CIN)

  - Five Outputs:  four summation outputs (S0 through S3) and a Carry output (COUT)

The pin diagrams (pinouts) of these two ICs are displayed in Figure 4-1.

Arithmetic circuits can be combinational arrays or individual ICs such as the 7483N or the 4008BT. In either case, the arithmetic is the same with the ICs being universal in their usage.

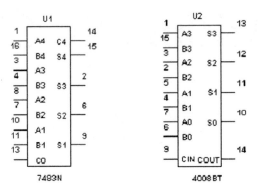

**Figure 4-1**   Pinouts for Full Adder ICs

## Activity 4.1: Adder Circuits

1. Basic adder circuits consist of **Half Adders** and **Full Adders**. The half adder is a circuit with two inputs, A and B, and two outputs, the summation output, S0, and the carry output, COUT. Half adders and full adders are generally constructed from basic gates rather than being manufactured as adder integrated circuits. The minimum size of an adder IC is the 4-bit adder as displayed in Figure 4-1.

2. Open circuits file **d04-01a**. This combinational adder circuit accomplishes the half adder function. Activate the switches and enter the resultant data in Table 4-1. Notice that a half adder circuit does not have a carry input from a previous stage.

| Half Adder Truth Table | | | |
|---|---|---|---|
| **Input B** | **Input A** | **Summation Output ($\Sigma$)** | **Carry Output** |
| 0 | 0 | | |
| 0 | 1 | | |
| 1 | 0 | | |
| 1 | 1 | | |

**Table 4-1**   Truth Table for Half Adder Circuit

3. A combinational full adder circuit is a little more complex than the half adder circuit with the addition of a carry input from a previous stage. Open circuits file **d04-01b**. Activate the switches and enter the results in Table 4-2.

| Full Adder Truth Table | | | | |
|---|---|---|---|---|
| Carry Input (Switch C) | Input B | Input A | Summation Output ($\Sigma$) | Carry Output |
| 0 | 0 | 0 | | |
| 0 | 0 | 1 | | |
| 0 | 1 | 0 | | |
| 0 | 1 | 1 | | |
| 1 | 0 | 0 | | |
| 1 | 0 | 1 | | |
| 1 | 1 | 0 | | |
| 1 | 1 | 1 | | |

**Table 4-2**   Truth Table for Full Adder

4. Open circuits file **d04-01c**. In this circuit, we will use the 4008 4-bit Binary Adder IC to check out the concept of four-place addition. The 4008BT is connected to add 6 to 5. The A inputs are connected for one number to be added, 0110 (6), and the B inputs for the other number, 0101 (5). The CIN input should be connected LOW. Activate the circuit. The Data Switch (J1) inputs the binary data into the adder. What are the sums of the addition? The binary sum is _____. The decimal sum is _____.

5. Open circuits file **d04-01d**. This 4008BT IC circuit is connected to add 2 (A) + 3 (B). Inputs are connected through the Data Switch J1. The A input that is connected to +5V (HIGH) is A _____ and the B inputs that are connected to +5 V (HIGH) are B _____ and B _____. What outputs are HIGH and what is the decimal sum? Outputs S _____ and S _____ are HIGH (ON), and the decimal sum of the addition process is _____.

6. Open circuits file **d04-01e**. This circuit is connected to add 5 (A inputs) + 2 (B input) + 1 (Carry input). Predict which output(s) will be on. Output(s) S _____ will be on. The decimal sum of the addition process is _____.

7. Open circuits file **d04-01f**. This circuit is connected to solve 8 (A inputs) + 9 (B inputs) + 1 (Carry input). Predict which output(s) will be on.

   Output(s) S _____ and C _____ will be on. The sum of

   the addition process is _____. This is the first circuit with a carry out (COUT).

8. Open circuits file **d04-01g**. Add the binary numbers 1111 (A) and 1111 (B) using this circuit (CIN is LOW). What is the decimal sum?

   The decimal sum is _____. What output light(s) are on?

   Output light(s) _____ is/are on.

## ● *Troubleshooting Problems:*

9. Open circuits file **d04-01h**. This circuit is supposed to add 7 (A inputs) and 5 (B inputs). The output sum appears to be wrong because S1

   remains on. Where and what is the problem? The problem is _____

   _____

   _____.

10. Open circuits file **d04-01i**. This circuit is supposed to add 7 (A inputs) + 5 (B inputs). The output sum appears to be wrong. Where and what is the

    problem? The problem is _____

    _____

    _____.

    This is a difficult trouble to pin down. There is an open input, and the best way to isolate the defective input is to disconnect HIGH inputs, and see if the output changes. If one of the HIGH inputs does not affect the output, then that input is probably open.

11. Open circuits file **d04-01j**. What are the input addends that this circuit

    is supposed to add? The A addend in decimal is _____. The B addend

    in decimal is _____. The output is obviously wrong, where and what

    is the problem? The problem is _____

    _____

    _____.

● *Circuit Construction:*

12. Open circuits file **d04-01k**. Using the components on the workspace, construct an adder circuit that will add the addends of A = 14, B = 12, and CIN = 1. The decimal sum should be _____ and outputs

S _____ and C _____ should be on.

## Activity 4.2: 1's (One's) Complement Subtractor Circuits

1. Sometimes addition is necessary and sometimes subtraction is necessary. It is easier to add than to subtract electronically, so a mathematical maneuver known as the **1's (one's) complement** is used for the subtraction process. The one's complement process essentially consists of three steps. First, the subtrahend (B input) is complemented (all of the ones and zeros that make up B are inverted). Secondly, this complement is added to A. Finally the sum is complemented again. If there is a carry out (COUT), a process entitled **end-around-carry** (EAC) is performed. If the answer to the problem turns out to be negative, then an indication has to be made that the answer is negative.

2. Open circuits file **d04-2a**. This complex circuit will add or subtract. Use the add/subtract switch to add and/or subtract the values indicated in Table 4-3 and enter your results. The first problem is given as an example.

| Values to be Added/Subtracted | | | | Result of Calculation | |
|---|---|---|---|---|---|
| A Inputs | Sign | B Inputs | | Sign | Value |
| 2 | + | 3 | = | + | 5 |
| 5 | − | 3 | = | | |
| 4 | − | 9 | = | | |
| 8 | + | 9 | = | | |
| 12 | − | 10 | = | | |
| 2 | − | 10 | = | | |
| 15 | − | 15 | = | | |
| 12 | + | 8 | = | | |
| 6 | − | 10 | = | | |
| 14 | + | 14 | = | | |

**Table 4-3**   1's Complement Adder/Subtractor

● *Troubleshooting Problem:*

3. Open circuits file **d04-02b.** There is a problem in this circuit. Add 8 + 10.

   What did you get for a sum? The decimal sum was _____ according to the output lamps. What should it have been? The decimal sum should

   have been _____ and output lamps _____ and _____ should

   have been on. What is wrong? The problem is _____

   _____

   _____ .

# Activity 4.3: Looking at the 2's (Two's) Complement Subtractor Circuit/Method

1. Two's (2's) complement subtraction of a number is similar to the one's complement method. First, the one's complement (of the B input) is accomplished; then one (1) is added to the result of the one's complement. This combination of one's complement plus 1 is designated as the two's complement. Finally, this two's complement result is added to the other number (the A input) which results in the answer to the original subtraction problem.

2. Open circuits file **d04-3a**. This circuit is constructed using the 74LS83N IC to subtract following the two's complement method. Use the circuit to subtract the values indicated in Table 4-4 on page 63 and enter the resultant data in the table. The first problem is given as an example. The circuit has a negative answer indicator. This indicator will be ON for a negative answer and OFF for a positive answer.

● *Troubleshooting Problem:*

3. Open circuits file **d04-03b**. In this hardwired (no switches) circuit, you are supposed to subtract 10 (B) from 8 (A), 8 − 10, with an expected answer of −2. What did you get for an answer? The

   answer was _____ according to the output lamps. There is a connection problem. The problem is _____

   _____

   _____ .

| Values to be Subtracted | | | | | | | | | | | | Result of Calculation | |
| A Inputs | | | | | | B Inputs | | | | | | | |
| Dec | A4 | A3 | A2 | A1 | Sign | Dec | B4 | B3 | B2 | B1 | | Sign | Value |
|---|---|---|---|---|---|---|---|---|---|---|---|---|---|
| 2 | 0 | 0 | 1 | 0 | – | 3 | 0 | 0 | 1 | 1 | = | – | 1 |
| 5 | 0 | 1 | 0 | 1 | – | 3 | 0 | 0 | 1 | 1 | = | + | |
| 4 | 0 | 1 | 0 | 0 | – | 9 | 1 | 0 | 0 | 1 | = | – | |
| 8 | 1 | 0 | 0 | 0 | – | 9 | 1 | 0 | 0 | 1 | = | – | |
| 12 | 1 | 1 | 0 | 0 | – | 10 | 1 | 0 | 1 | 0 | = | + | |
| 2 | 0 | 0 | 1 | 0 | – | 19 | 1 | 0 | 1 | 0 | = | – | |
| 15 | 1 | 1 | 1 | 1 | – | 15 | 1 | 1 | 1 | 1 | = | + | |
| 12 | 1 | 0 | 1 | 0 | – | 8 | 1 | 0 | 0 | 0 | = | + | |
| 6 | 0 | 1 | 1 | 0 | – | 10 | 1 | 0 | 1 | 0 | = | – | |
| 14 | 1 | 1 | 1 | 0 | – | 14 | 1 | 1 | 1 | 0 | = | + | |

**Table 4-4**    Using the 2's (Two's) Complement Subtractor Circuit

## ● *Circuit Construction:*

4. Open circuits file **d04-03c**. Using the components on the workspace, construct a hard-wired subtractor circuit that will subtract 9 from 6. Draw your constructed circuit on Figure 4-2. What was the answer from the subtraction process? The answer to the subtraction problem

was _____.

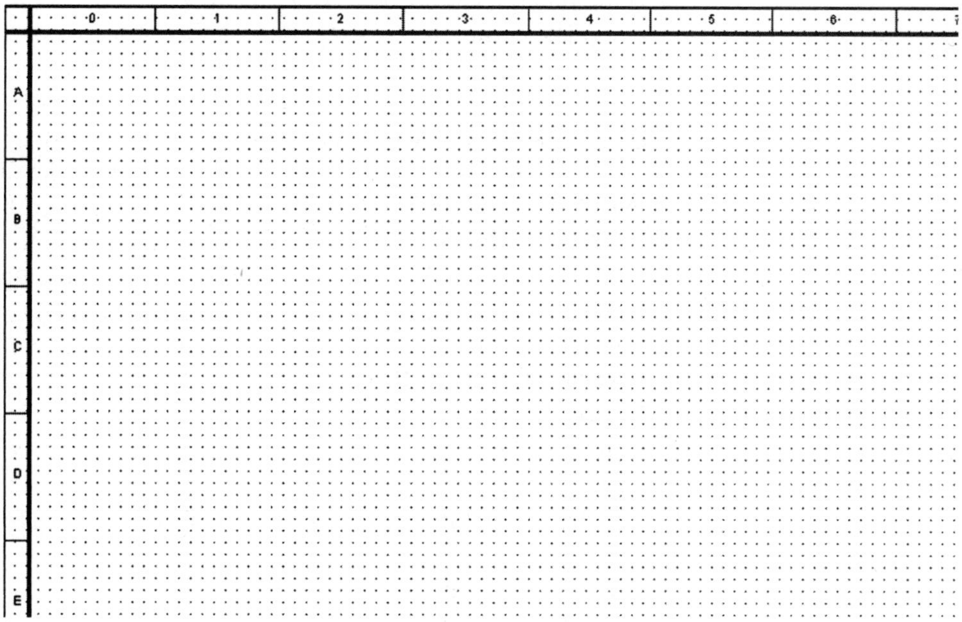

**Figure 4-2**    Constructed Subtractor Circuit

# 5. Open Collector Gates and Tri-State Circuits

**References**

*MultiSIM* 2001

*MultiSIM* 2001 Study Guide

**Objectives**   After completing this chapter, you should be able to:

- Use open-collector gates.
- Use circuits with tri-state buffers and inverters.
- Determine logic for complex tri-state circuits.
- Construct open collector circuits.
- Construct tri-state circuits.
- Troubleshoot open-collector and tri-state logic circuits.

## Introduction

One disadvantage of TTL gates, and also a problem with other families of ICs, is a potential current problem when output circuits are connected to a common connection such as an address or data bus in microprocessor circuits. The common connection of outputs of logic gates is sometimes designated as a "wired AND". The problem occurs when one or more of the wired output circuits turns ON (goes HIGH), and the others stay OFF (LOW). The device that turns ON will sink (draw) current from all of the OFF devices and allow enough current to flow to ultimately destroy the ON device. This problem can be overcome by the use of open collector and tri-state circuits.

**Open collector** circuits are circuits that are able to operate at normal IC voltages such as +5 V and are also able to drive an output load connected to a much higher voltage. Open collector (OC) output capabilities are provided for all basic IC logic gates. Some examples are the 7403 Quad Two-input NAND (OC), the 7405 Hex Inverter (OC), and the 7422 Dual Four-input NAND. These examples are displayed in Figure 5-1. The open collector gate technique depends on an external **pull-up** resistor to provide a current path during the ON condition rather than an internal output transistor within the IC. The internal resistor is missing in OC integrated circuits.

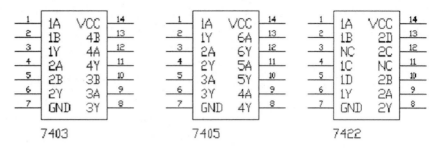

**Figure 5-1**    Some Pinouts of ICs with Open Collector Outputs

**Tri-state** gates are buffer and inverter circuits that have three possible output conditions. These conditions are the normal HIGH and LOW logic states and a third state, the **high impedance state**, occurring when the output has no effect on any circuits that are connected to it (this could be called a floating state). Some examples of tri-state gate circuits are the 74LS240N Tri-State Buffer Inverter, the 74LS241N Tri-State Buffer (non-inverting), and the 74LS245N Octal Bus Transceiver displayed in Figure 5-2. These tri-state circuits prevent the problem of excessive currents by specifically turning ON only the addressed circuit and thereby inhibiting any current flow participation by the non-addressed circuits, holding them in a HIGH impedance state (floating). This prevents the OFF circuits from sending unwanted and excessive currents through the ON circuit.

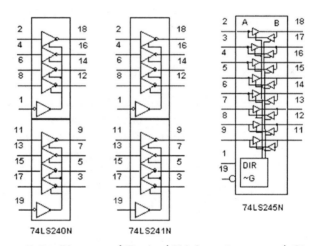

**Figure 5-2**    Pinouts of Typical Tri-State Integrated Circuits

## Activity 5.1:  Open Collector Outputs

1. Open collector (OC) outputs are usually used to float (HIGH impedance -Z) the output pins of ICs that are connected to a common connection point such as a microprocessor bus. ICs not having open collector outputs (or tri-state capabilities) provide binary ones and zeros (HIGHs and LOWs) at their output terminals. A common connection point with several connected circuits trying to go HIGH or LOW simultaneously, can create excessive currents through the internal circuitry of the ICs. This condition can be prevented with the use of open collector outputs. The

open collector circuit floats the output of any IC that is not being addressed or told to go HIGH or LOW. If confusion (an improper logic condition) occurs with several ICs going HIGH and LOW simultaneously, then the external pull-up resistor will handle the excessive current and there will be no harm to the connected ICs.

2. Open circuits file **d05-01a**. Actuate the input switches for this 74LS03N (OC) NAND gate and enter the resultant data in Table 5-1. Now, use S2 to connect the pull-up resistor R1 to TPA and re-actuate the switches. Enter this additional data in Table 5-1.

| Open Collector Pull-up Resistor Not Connected S2 Open | | | Hot Collector Pull-up Resistor Connected S2 Closed | | |
|---|---|---|---|---|---|
| **B** | **A** | **Y** | **B** | **A** | **Y** |
| 0 | 0 | | 0 | 0 | |
| 0 | 1 | | 0 | 1 | |
| 1 | 0 | | 1 | 0 | |
| 1 | 1 | | 1 | 1 | |

**Table 5-1**   74LS03N (OC) NAND Gate Truth Table

3. Open circuits file **d05-01b**. Actuate the input switch for this 74LS05N (OC) Inverter and enter the resultant data in Table 5-2.

| Open Collector Pull-up Resistor Not Connected S2 Open | | Hot Collector Pull-up Resistor Connected S2 Closed | |
|---|---|---|---|
| **A** | **Y** | **A** | **Y** |
| 0 | | 0 | |
| 1 | | 1 | |

**Table 5-2**   Inverter (OC) Truth Table

4. Open circuits file **d05-01c**. Actuate the input switches for this 74LS22N (OC) Dual Four-input NAND gate and enter the resultant data in Table 5-3.

| Open Collector Pull-up Resistor Not Connected S2 Open | | | | | Hot Collector Pull-Up Resistor Connected S2 Closed | | | | |
|---|---|---|---|---|---|---|---|---|---|
| **D** | **C** | **B** | **A** | **Y** | **D** | **C** | **B** | **A** | **Y** |
| 0 | 0 | 0 | 0 | | 0 | 0 | 0 | 0 | |
| 0 | 0 | 0 | 1 | | 0 | 0 | 0 | 1 | |
| 0 | 0 | 1 | 0 | | 0 | 0 | 1 | 0 | |
| 0 | 0 | 1 | 1 | | 0 | 0 | 1 | 1 | |
| 0 | 1 | 0 | 0 | | 0 | 1 | 0 | 0 | |
| 0 | 1 | 0 | 1 | | 0 | 1 | 0 | 1 | |
| 0 | 1 | 1 | 0 | | 0 | 1 | 1 | 0 | |
| 0 | 1 | 1 | 1 | | 0 | 1 | 1 | 1 | |
| 1 | 0 | 0 | 0 | | 1 | 0 | 0 | 0 | |
| 1 | 0 | 0 | 1 | | 1 | 0 | 0 | 1 | |
| 1 | 0 | 1 | 0 | | 1 | 0 | 1 | 0 | |
| 1 | 0 | 1 | 1 | | 1 | 0 | 1 | 1 | |
| 1 | 1 | 0 | 0 | | 1 | 1 | 0 | 0 | |
| 1 | 1 | 0 | 1 | | 1 | 1 | 0 | 1 | |
| 1 | 1 | 1 | 0 | | 1 | 1 | 1 | 0 | |
| 1 | 1 | 1 | 1 | | 1 | 1 | 1 | 1 | |

**Table 5-3**   Truth Table for Four-Input NAND Logic Gate

## ● *Troubleshooting Problems:*

5. Open circuits file **d05-01d**. This circuit has a component problem; the output does not go HIGH when the open collector is connected to a voltage source. What is the problem? The problem is _____

_____

_____.

6. Open circuits file **d05-01e**. This circuit is not working correctly; it has a connection problem. The output doesn't change at all. What is the problem? The problem is _____

_____

_____.

## ● *Circuit Construction:*

7. Open circuits file **d05-01f** and construct a two-input XOR circuit using the components on the workspace. The Boolean equation to turn on the lamp is $A + \overline{B} = Y$. Activate the circuit and check its operation. Did the

   output lamp, X1, turn on, Yes _____ or No _____?

## Activity 5.2: Tri-State Circuits

1. There are many types of Tri-State devices ranging from simple logic inverter/buffer gates to very complex tri-state bus control circuits. A bus is a common connection point used to distribute signals to various locations within electronic equipment and especially in microprocessors. In the following circuits, the control signals route data to the bus through the tri-state devices connected to the bus. The term tri-state (three state) means that this category of circuits has three possible output conditions. The three possible output conditions are HIGH output, LOW output, and high impedance (floating) output. The symbols for two basic logic gates are displayed in Figure 5-3, a Buffer and an Inverter. Both of these circuits are shown with an **Active HIGH** for the control signal. The control signal connection point is the vertical line at the top of the tri-state integrated circuits.

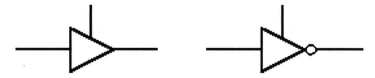

**Figure 5-3**   Tri-State Logic Gates with 'Active HIGH' Control

2. Figure 5-4 displays circuits that use **Active LOW** control. Notice the circles on the control inputs at the top of the ICs. These circles indicate that an active LOW input is required and when this LOW is supplied, the input logic signals are allowed to flow from the input terminal to the output terminal. In this case, if the control terminal were placed HIGH, the output terminal would be in the HIGH impedance state (floating).

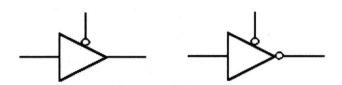

**Figure 5-4**   Tri-State Logic Gates with 'Active LOW' Control

3. Open circuits file **05-02a**. In this circuit, using 74LS240N Tri-State Buffer Inverters, the circuit is switched between two square wave inputs, one at 1000 Hz and the other at 2000 Hz. The control switch determines which one of the two 74LS240N integrated circuits is enabled to pass its input square wave signal to the bus (the common connection point to the right of the two ICs). If the control switch is up (+5 V), circuit

U _____ allows signals to flow to the output bus and if the control

switch is down (GND), U _____ allows signals to flow to the output bus. The ICs designated IC3, IC4, and IC5 are examples of circuits connected to a common bus. Draw the two waveforms (1000 Hz and 2000 Hz) in their respective locations on Figure 5-5. Ignore the +5 V signal (a straight line) displayed on the oscilloscope when one of the data waveforms are present, just draw the data waveforms in their respective top and bottom locations.

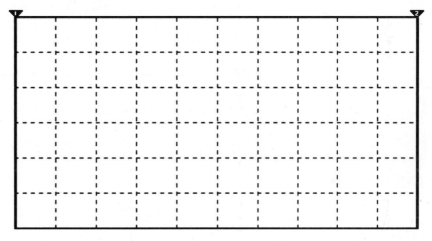

**Figure 5-5**   Bus Waveforms Passed Through 74LS240N ICs

4. Gates are the basic tri-state devices, but there are many other types and much more complicated circuits that fall into the tri-state category. Generally all of these circuits are used to communicate microprocessor data and address and control information using the internal microprocessor buses. For example, the 74LS245N Octal Bus Transceiver provides a means where data can be transferred bidirectionally using an internal direction circuit that steers the data from A to B or from B to A. This bidirectional circuit places one of the two buses in the receive mode (HIGH). When one of these buses is HIGH, the tri-state circuits connected to that bus are enabled to pass data, the bus itself becomes the control line rather than the data handling connection point. The 74LS245N is commonly used for read/write situations where data is being written to a memory location or is being transferred out of a memory location in microprocessor equipment. There are many other specialized tri-state circuits that are used for similar microprocessor applications.

5. Open circuits file **05-02b**. In this circuit, again using the 74LS245N Octal Bus Transceiver, the data passes from left to right (A-side to the B-side) when the switch is in the up position (+5 V) and from right to left (B-side to the A-side) when the switch is in the down position (GND). With the switch in the up position, the data (1000 HZ) flows into the bus

through Pin _____ and out of the bus through Pin _____ to Channel A on the oscilloscope. With the switch in the down position, the

data (2000 Hz) flows in on Pin _____ and out on Pin _____ to Channel B on the oscilloscope. Pin G on the IC is the **enable** pin and has to be connected LOW (GND) for the data to pass through the IC. The small circle on Pin 19 (~G), the enable input, indicates that the input is an active LOW, the input has to be LOW for the circuit to do its function, in this case to pass data from one side of the IC to the other.

6. Draw the two waveforms (1000 Hz and 2000 Hz) in their respective locations on Figure 5-6. Ignore the +5 V signal (a straight line) displayed on the oscilloscope when one of the data waveforms are present, just draw the data waveforms in their respective top and bottom locations.

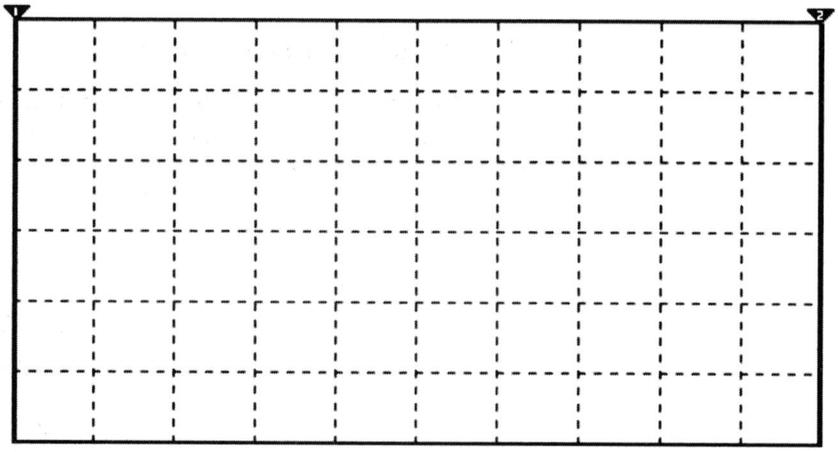

**Figure 5-6**   More Bus Waveforms Passed Through 74LS240N ICs

## ● *Troubleshooting Problems:*

7. Open circuits file **05-02c**. In this circuit, using the 74LS245N Octal Bus Transceiver, no data is passing through the circuit; it has a connection

problem. What is wrong? The problem is  _____

_____

_____ .

Check the previous circuit of Step 5 for proper connections.

8. Open circuits file **05-02d**. In this circuit using the 74LS240N Tri-State Inverter Buffer, 1000 Hz and 2000 Hz data is supposed to be placed on the bus depending on which half of the circuit is enabled. The oscilloscope indicates that some of the data is not getting to the bus. Use the DMM and the oscilloscope to locate the problem. The problem is _____

_____

_____.

9. Open circuits file **05-02e**. This circuit uses the 74LS366N Tri-State Hex Bus Driver. This IC provides six inverted outputs. The ~G1 *and* ~G2 inputs are the enable inputs for all of the inverters. There is a problem with this circuit. Toggle the switch S1 and observe the output lamps. They should all turn on and off. What is wrong? The problem is _____

_____

_____.

● *Circuit Construction:*

10. Open circuits file **d05-02f** and construct a circuit similar to the circuit of Figure 5-7 using the components on the workspace.

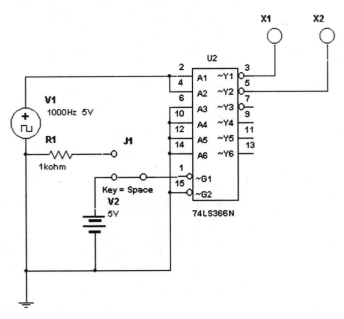

**Figure 5-7**   Construction of Circuit with the 74LS366N IC

# 6. Sequential Circuits: NAND/NOR Latches

> **References**
> *MultiSIM* 2001
> *MultiSIM* 2001 Study Guide

**Objectives**  After completing this chapter, you should be able to:

- Determine input and output states of NAND/NOR latch circuits.
- Construct S-R latch circuits with NAND and NOR gates.
- Construct and use Gated S-R latch circuits.
- Construct and use transparent D latch circuits.
- Troubleshoot typical latch circuits.

**Latch** circuits are sequential circuits with two inputs, called **SET** and **RESET**, which enable a circuit to store a logic 0 (representing RESET) or a logic 1 (representing SET) until a later input brings about a change in the set/reset logic state. Some latch circuits can have additional Enable input. When we refer to the condition of the circuit as being SET or RESET, we are referring to the condition of the $Q$ output as displayed in the typical latch circuit of Figure 6-1. These terms can be further defined as:

- **SET**
   1. A HIGH (a logic one) condition stored in a latch circuit.
   2. An input that causes the latch circuit to store a HIGH (a logic one).

- **RESET**
   1. A LOW (a logic zero) condition stored in a latch circuit.
   2. An input that causes the latch circuit to store a LOW (a logic zero).

- **ENABLE**
   1. An active input to a latch circuit that allows it to change states.
   2. A synchronizing signal.

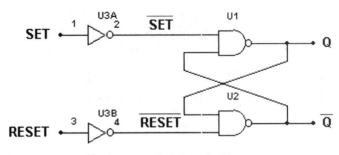

**Figure 6-1**  S-R Latch Circuit

These circuits can be further characterized as circuits that have two outputs that are always digital opposites or complements of each other. The two outputs $Q$ and $\overline{Q}$, assume opposite states. If $Q$ is LOW (0 V), then $\overline{Q}$ is HIGH (+5 V), and the device is said to be LOW or in the zero state. The state of the device is always indicated by the HIGH or LOW state of the $Q$ output. If $Q$ is HIGH (+5 V), then $\overline{Q}$ is LOW (0 V), and the state of the device is said to be HIGH or in the one state. By having the ability to retain a logic state, latch circuits can store data and provide service as memory devices.

## Activity 6.1:  The $\overline{\text{SET-RESET}}$ (S-R) Latch Circuit (Crossed-NAND)

1. The $\overline{\text{SET-RESET}}$ **(S-R) Latch** circuit is also known as a complementary NAND or a crossed-NAND Latch circuit (NOR ICs can be used for this purpose also). It is the simplest form of latch and is seldom used in this simple form. The only common usage of the S-R latch is to debounce input switch closures. All flip-flop circuits, no matter how complex, have one or more S-R latches in their interior circuitry. S-R latches are in the center of every one of the more complex flip-flops that we will be studying in the next chapter. A truth table for an S-R latch is displayed in Figure 6-2. Of the four conditions displayed, the first one (1-1) is prohibited according to manufacturers' literature and generally is not supposed to be used. The fourth condition (0-0) holds the latch in its present state at that specific time, either $\overline{\text{SET}}$ or $\overline{\text{RESET}}$ ($Q$ or $\overline{Q}$).

| SET | RESET | $\overline{\text{SET}}$ | $\overline{\text{RESET}}$ | $Q$ | $\overline{Q}$ | CONDITION |
|---|---|---|---|---|---|---|
| 1 | 1 | 0 | 0 | 1 | 1 | **Unused State** |
| 1 | 0 | 0 | 1 | | | **Set** |
| 0 | 1 | 1 | 0 | | | **Reset** |
| 0 | 0 | 1 | 1 | | | **Unchanged State** |

**Figure 6-2**  Truth Table for S-R Latch Circuit (NAND)

2. Open circuits file **d06-01a**. This circuit, similar to the one displayed in Figure 6-1, is an S-R latch. Activate the switches and verify the truth table of Figure 6-2. Enter your results into Table 6-1. Switch A is the SET switch and switch B is the RESET switch.

| SET | RESET | $\overline{SET}$ | $\overline{RESET}$ | $Q$ | $\overline{Q}$ | CONDITION |
|-----|-------|------|--------|-----|-----|-----------|
| 1 | 1 | 0 | 0 | 1 | 1 | **Unused State** |
| 1 | 0 | 0 | 1 | | | **Set** |
| 0 | 1 | 1 | 0 | | | **Reset** |
| 0 | 0 | 1 | 1 | | | **Unchanged State** |

**Table 6-1**  Truth Table for S-R Latch Circuit

### ● *Troubleshooting Problem:*

3. Open circuits file **d06-01b**. In this S-R latch circuit the $Q$ output remains HIGH. What is the problem? The problem is _____

_____

_____.

### ● *Circuit Construction:*

4. Open circuits file **d06-01c**. Using the components on the workspace, construct a circuit similar to Figure 6-3.

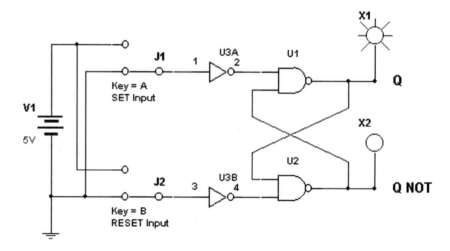

**Figure 6-3**  Constructed S-R Latch Circuit

## Activity 6.2: The SET-RESET (S-R) Latch Circuit (Crossed-NOR)

1. In this activity, a SET-RESET (S-R) Latch uses an alternative method, NOR gates rather than NAND gates, to develop the circuit. This circuit is also known as a complementary NOR or a crossed-NOR latch circuit. Again, it is a very simple form of latch and is seldom used in this simple form except as a debounce circuit. The truth table for an S-R latch (using crossed NORs) is displayed in Figure 6-4. Of the four conditions displayed, the first one (1-1) is prohibited according to some literature and generally is not used. The fourth condition (0-0) holds the latch in its present state at that time, either $\overline{SET}$ or $\overline{RESET}$.

| SET | RESET | $\overline{SET}$ | $\overline{RESET}$ | $Q$ | $\overline{Q}$ | CONDITION |
|-----|-------|------|--------|-----|-----|-----------|
| 1 | 1 | 0 | 0 | 0 | 0 | **Unused State** |
| 1 | 0 | 0 | 1 | 1 | 0 | **Set** |
| 0 | 1 | 1 | 0 | 0 | 1 | **Reset** |
| 0 | 0 | 1 | 1 | $Q$ | $\overline{Q}$ | **Unchanged State** |

**Figure 6-4**  Truth Table for S-R Latch Circuit (NOR)

2. Open circuits file **d06-02a**. The circuit, similar to the one displayed in Figure 6-4, is an S-R latch circuit composed of NOR gates instead of NAND gates. Actuate the switches and verify the truth table of Figure 6-5. Enter your resultant data into Table 6-2. Notice the change in location of the $Q$ and $\overline{Q}$ ($Q$ NOT) outputs.

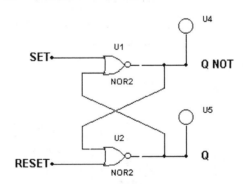

**Figure 6-5**  S-R Latch Circuit (NOR)

| SET | RESET | $Q$ | $\overline{Q}$ | CONDITION |
|-----|-------|-----|-----|-----------|
| 1 | 1 | 0 | 0 | **Unused State** |
| 1 | 0 | | | **Set** |
| 0 | 1 | | | **Reset** |
| 0 | 0 | | | **Unchanged State** |

**Table 6-2**  Truth Table for S-R Latch Circuit (NOR)

## ● *Troubleshooting Problems:*

3. Open circuits file **d06-02b**. There is a problem in this Crossed-NOR

   S-R latch circuit; the $Q$ output remains HIGH. The problem is _____

   _____

   _____.

4. At this point, you should have noticed that there are several differences between the two types of S-R latch circuits that we have been studying, the crossed-NAND and the crossed-NOR. One difference is the value of $Q$ and $\overline{Q}$ in the unused state, and the other is that the crossed-NOR latch requires active HIGH inputs and the crossed-NAND requires active LOW inputs. The crossed-NOR will change its state when its input goes HIGH, and the crossed-NAND will change its state when input goes LOW.

## ● *Circuit Construction:*

5. Open circuits file **d06-02c**. Using the components on the workspace, construct a circuit similar to the circuit of Step 3. Draw your circuit on Figure 6-6.

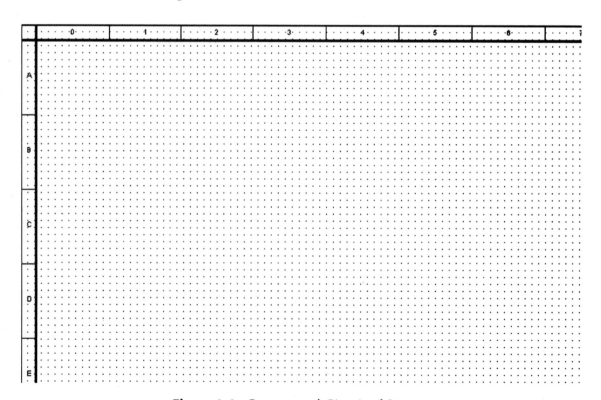

**Figure 6-6**  Constructed Circuit of Step 6

## Activity 6.3:  Using an S-R Latch Circuit to Debounce a Switch

1. The metal contacts of mechanical switching devices tend to bounce when they are actuated. This bouncing condition can cause a digital circuit to react to each bounce as a signal, treating each bounce as an individual input. This could result in a digital circuit changing states several times with only intentional input. A typical debounce circuit uses the $\overline{\text{SET}}$-$\overline{\text{RESET}}$ (S-R) Latch to solve the bouncing problem. When the switch is actuated (switches), the latch circuit changes its state once and remains in that changed state even though the switch bounces, causing the input signal to go HIGH and LOW a number of times. Figure 6-7 displays a typical debounce circuit using NAND gates.

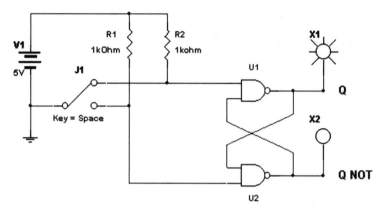

**Figure 6-7**  A Typical Debounce Circuit

2. Open circuits file **d06-03a**. This circuit is designed to debounce switch S1 (A). Switch S2 (spacebar) is installed in the circuit to simulate contact bounce after switch S1 is closed. First move switch S1 to the down position. What happens? The _____ output goes HIGH.

   - Toggle switch S2 several times simulating contact bounce. What happens?_____

     _____.

   - Close switch S2. Move switch S1 to the upper position. What happened? The _____ output goes HIGH.

   - Now toggle switch S2 several times simulating contact bounce. What happens?_____

     _____.

3. In the previous steps, switch S2 is simulating contact bounce after the initial switch closure (switch S1) occurs. The outputs after the S1 closure should not have changed when switch S2 was toggled.

## ● *Circuit Construction:*

4. Open circuits file **d06-03c**. Using the components on the workspace, construct a circuit similar to the circuit of Figure 6-7. Draw your circuit on Figure 6-8.

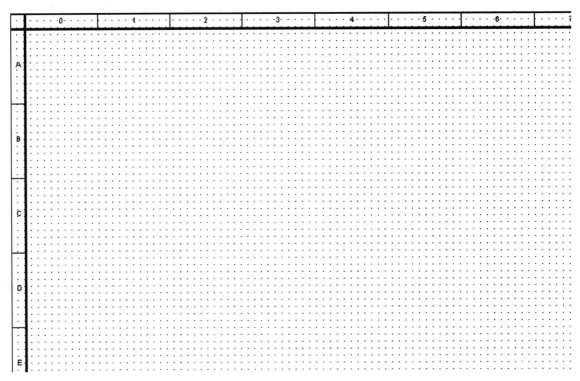

**Figure 6-8** Constructed Circuit of Step 4

## Activity 6.4: The Gated S-R Latch Circuit

1. The **Gated S-R Latch (NAND)** circuit uses additional NAND gates to control the actuation of the latch circuit with an Enable input. The S-R latch circuit will not change its state unless the Enable input is present. As can be seen by the logic of Figure 6-9, when the Enable input is zero (LOW), the latch cannot change states and when the Enable input is one (HIGH), the latch is enabled to change its logic state. When an Enable input pin is part of an integrated circuit, the output condition will not change when the Enable signal is logically missing. Figure 6-10 displays a typical truth table for an S-R latch circuit.

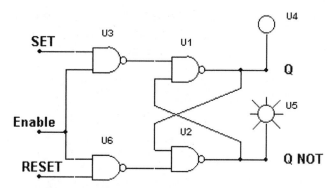

**Figure 6-9**  Gated S-R Latch Circuit

| Enable | SET | RESET | $Q$ | $\overline{Q}$ | |
|--------|-----|-------|-----|-----------------|---|
| 0 | 0 | 0 | $Q$ | $\overline{Q}$ | |
| 0 | 0 | 1 | $Q$ | $\overline{Q}$ | |
| 0 | 1 | 0 | $Q$ | $\overline{Q}$ | Unchanged State |
| 0 | 1 | 1 | $Q$ | $\overline{Q}$ | |
| 1 | 0 | 0 | $Q$ | $\overline{Q}$ | |
| 1 | 0 | 1 | 0 | 1 | |
| 1 | 1 | 0 | 1 | 0 | |
| 1 | 1 | 1 | 1 | 1 | Unused State |

**Figure 6-10**  Truth Table for a S-R Gated Latch Circuit

2. Open circuits file **d06-04a**. This circuit is a typical gated S-R latch circuit using a switch (S2) to provide an Enable signal similar to the circuit of Figure 6-9. Actuate the switches and enter the resulting data in Table 6-3.

| Enable | SET | RESET | $Q$ | $\overline{Q}$ | |
|--------|-----|-------|-----|-----------------|---|
| 0 | 0 | 0 | | | |
| 0 | 0 | 1 | | | |
| 0 | 1 | 0 | | | Unchanged State |
| 0 | 1 | 1 | | | |
| 1 | 0 | 0 | | | |
| 1 | 0 | 1 | | | |
| 1 | 1 | 0 | | | |
| 1 | 1 | 1 | 1 | 1 | Unused State |

**Table 6-3**  Truth Table for Gated S-R Latch Circuit

● *Troubleshooting Problem:*

3. Open circuits file **d06-04b**. There is a problem in this S-R latch circuit; the $Q$ (X1) output remains LOW. The problem is _____

_____

_____ .

● *Circuit Construction:*

4. Open circuits file **d06-04c**. Using the components on the workspace, construct a circuit similar to the circuit of Figure 6-9 and the MultiSIM circuit of Step 2. Draw your circuit on Figure 6-11.

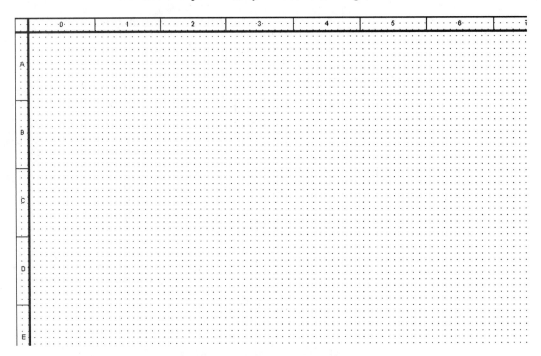

**Figure 6-11** Constructed Circuit of Step 4

## Activity 6.5: The Gated D (Transparent) Latch Circuit

1. The **Gated D (Transparent) Latch** circuit is used to overcome problems that are endemic with the basic S-R latch circuit. The possibility of a HIGH simultaneously on both inputs of the S-R latch circuit (the unused state) is overcome by using a NOT circuit to invert the input applied to the SET terminal and then applying that inverted input to the RESET terminal. In that manner, the two inputs can never be HIGH at the same time. This advantage is joined with the additional benefit of only one input being required to SET or RESET the latch circuit. A typical Transparent D Latch circuit is displayed in Figure 6-12. The term **transparent** indicates the ability of a latch circuit to follow (and store)

the input level when the Enable input is active (in this case, HIGH). By the way, the D stands for data.

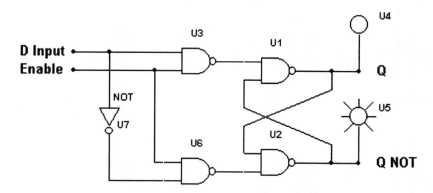

**Figure 6-12**   The D Latch Circuit

2. Open circuits file **d06-05a**. Develop truth table information for this circuit and enter the resultant data in Table 6-4.

| Enable | D Input | $Q$ | $\overline{Q}$ | |
|--------|---------|-----|----------------|---|
| 0 | 0 | | | **Unchanged State** |
| 0 | 1 | | | |
| 1 | 0 | | | |
| 1 | 1 | | | |

**Table 6-4**   Truth Table for a D Latch Circuit

## ● *Troubleshooting Problem:*

3. Open circuits file **d06-05b**. There is a problem with this circuit; what is it? The problem is _____

_____

_____.

## ● *Circuit Construction:*

4. Open circuits file **d06-05c**. Using the components on the workspace, construct a circuit similar to the circuit of Figure 6-12. Draw your circuit on Figure 6-13 on page 82.

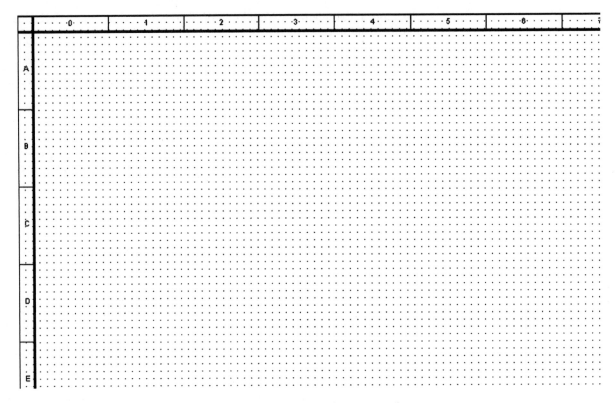

**Figure 6-13** Constructed Circuit of Step 4

## Activity 6.6: The 74LS75N Quad S-R Latch IC

1. The 74LS75N Quad S-R Latch IC is an integrated circuit that contains four separate S-R latch circuits and an Enable input for control.

2. Open circuits file **d06-06a**. This circuit is designed to alternate two sections of the latch circuit outputs between SET and RESET. Actuate the switches and enter the resultant data into Table 6-5. Note that all four of the available latch circuits are being used in this circuit. Notice that if the enable switch is left HIGH (+5 V), the latch outputs will follow the data switch.

| Enable | D Input | $1Q$ | $\overline{1Q}$ | $2Q$ | $\overline{2Q}$ | $3Q$ | $\overline{3Q}$ | $4Q$ | $\overline{4Q}$ | |
|--------|---------|------|------|------|------|------|------|------|------|---|
| 0 | 0 | | | | | | | | | **Unchanged State** |
| 0 | 1 | | | | | | | | | |
| 1 | 0 | | | | | | | | | **Data State** |
| 1 | 1 | | | | | | | | | |

**Table 6-5**  74LS75N Latch Circuit Truth Table

## ● *Troubleshooting Problem:*

3. Open circuits file **d06-06b**. This circuit is the same as the circuit of Step 2 except that only two outputs are being used. Outputs $2Q$ and $\sim2Q$ are

not changing. What is wrong? The problem is _____

_____

_____.

### ● *Circuit Construction:*

4. Open circuits file **d06-06c**. Using the components on the workspace, construct a circuit similar to the circuit of Figure 6-14. Draw your circuit on Figure 6-15.

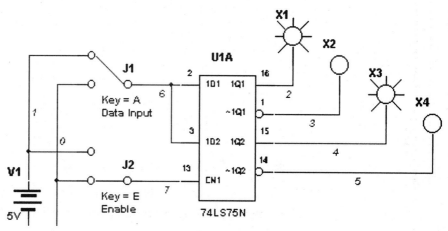

**Figure 6-14**  The 74LS75N Quad S-R Latch IC (Section 1)

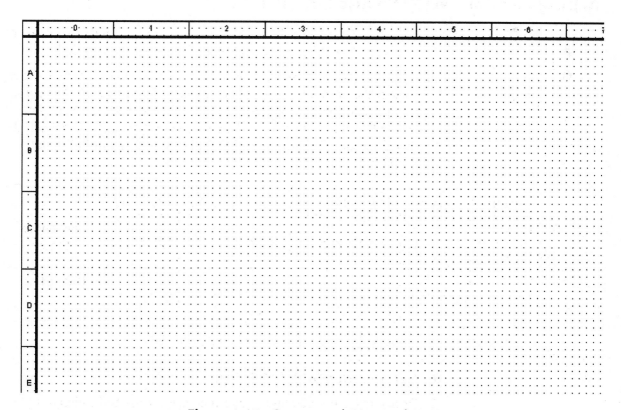

**Figure 6-15**  Constructed Circuit of Step 4

## Activity 6.7: Using the 4043BD Quad (CMOS) S-R Latch IC

1. The **4043 Quad (CMOS) S-R Latch IC** is a CMOS integrated circuit that contains four separate S-R latch circuits and an Enable input for control.

2. Open circuits file **d06-07a**. This circuit is designed to alternate all four sections of the IC between SET and RESET as a group. Actuate the switches and enter the resultant data into Table 6-6.

| Enable | D Input | O0 | O1 | O2 | O3 | |
|--------|---------|----|----|----|----|--|
| 0 | 0 | | | | | **Unchanged State** |
| 0 | 1 | | | | | |
| 1 | 0 | | | | | **Data State** |
| 1 | 1 | | | | | |

**Table 6-6**   4043BD Quad (CMOS) Latch IC Truth Table

### ● *Troubleshooting Problem:*

3. Open circuits file **d06-07b**. This circuit is the same as the circuit of Step 2 except that none of the outputs change when the input changes. What

   is wrong? The problem is _____

   _____

   _____.

### ● *Circuit Construction:*

4. Open circuits file **d06-07c**. Using the components on the workspace, construct a circuit similar to the circuit of Figure 6-16. Draw your circuit on Figure 6-17.

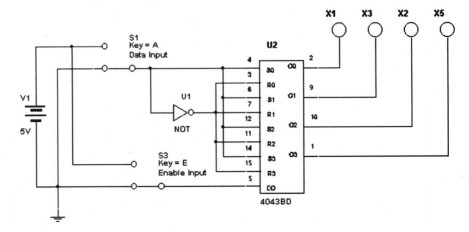

**Figure 6-16**   Latching Circuit Using the 4043BD Quad S-R latch IC

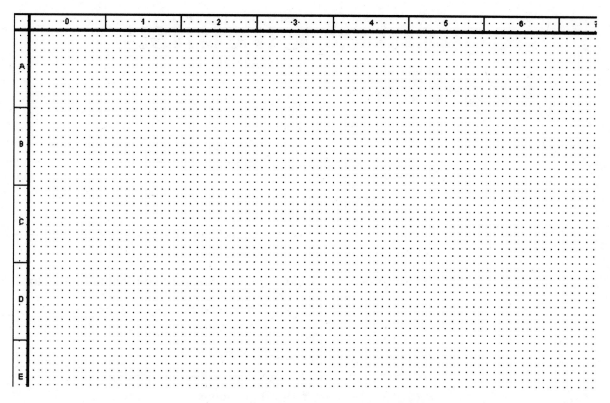

**Figure 6-17**  Constructed Circuit Using the 4043BD Quad Latch

# 7. Sequential Circuits: Flip-Flops

| References |
| --- |
| *MultiSIM* 2001 |
| *MultiSIM* 2001 Study Guide |

**Objectives**  After completing this chapter, you should be able to:

- Determine input and output states of flip-flop circuits.
- Construct flip-flop circuits with NAND gates.
- Construct and use D flip-flop circuits.
- Construct and use J-K flip-flop circuits.
- Construct and use T flip-flop circuits.
- Troubleshoot typical flip-flop circuits.

**Flip-Flop** circuits, along with latch circuits, fall into the category of sequential circuits. Sequential circuits change states according to present inputs in relationship to the historical state of the circuit. Within the category of sequential circuits, a flip-flop can be further defined as:

- A sequential circuit based on a latch whose output changes when its **CLOCK** input receives either an edge or a pulse, depending on the device.

- A circuit that typically has an edge detector (CLOCK) input that converts the active edge (leading or trailing) of an input CLOCK signal into an actuating pulse at the internal gates of the latch circuit. This actuating pulse is used to synchronize the SET and RESET inputs.

## Activity 7.1:  The Logic of an Edge-Triggering Circuit

1. The purpose of an **Edge-Triggering** circuit is to turn the leading or trailing edge of a CLOCK signal into a short duration pulse. This pulse will be used to synchronize the SET and RESET inputs of a typical flip-flop circuit. A typical edge triggering logic circuit is displayed in Figure 7-1.

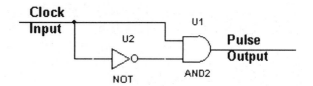

**Figure 7-1**  An Edge-Triggering Circuit

2. The duration of the output pulse of an edge-triggering circuit is determined by the **propagation delay** of the NOT portion of the above circuit. Initially (assumed for the sake of developing the theory of the circuit), the upper input to the AND2 gate is LOW and the lower input would thus be HIGH because of the inverting function of the NOT gate. The leading edge of the positive-going clock pulse forces the upper input HIGH while the lower input remains HIGH. The lower input will not go LOW (the delay factor) until the period of time required for the output of the NOT gate to change to the new (LOW) condition has passed. That period of time is entitled "propagation delay." The result is a pulse with a duration equal to the period of propagation delay.

3. Open circuits file **d07-01**. In this project you are to determine the duration of the output pulse developed by the circuit. The input square wave is set for 10 MHz. Use the oscilloscope to view the waveform. When you activate the oscilloscope, you will see the pulses running across the scope face. Go to the lower right corner of the scope (trigger area) and push the **Sing**. button. This should stop the signal and display one pulse at the left side of the scope face. Use the cursor to measure the duration of the pulse. What is the duration of the output pulse? The output pulse is about

_____ nanoseconds in duration.

## Activity 7.2: The Edge-Triggered D Flip-Flop

1. The **Edge-Triggered D Flip-Flop** circuit depends on the edge of the input CLOCK signal to synchronize a change in its SET or RESET condition. Figure 7-2 displays the 74LS174N Hex D Flip-Flop with CLEAR. Notice that there are six $Q$ outputs and no $\overline{Q}$ outputs.

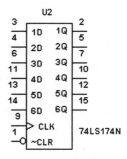

**Figure 7-2**  74LS174N Hex D Flip-Flop

2. Open circuits file **d07-02a**. This circuit uses the 7474N Dual D Flip-Flop. This IC has Clear and Preset inputs along with Data and Clock inputs. There are also two outputs for each IC ($Q$ and $\overline{Q}$). Activate the switch and note that the $Q$ and $\overline{Q}$ outputs revert to the opposite state shortly after the switch is activated. Fill Table 7-1 with the data from your observations.

| Switch Inputs | Outputs | | | |
|---|---|---|---|---|
| | 1Q | ~1Q | 2Q | ~2Q |
| 0 V (LOW) | | | | |
| +5 V (HIGH) | | | | |

**Table 7-1**   7474N Dual D Flip-Flop Data

3. Use the oscilloscope to determine the frequency of the input CLOCK signal coming from the function generator? The frequency of the

   CLOCK signal is _____ Hz.

● *Troubleshooting Problems:*

4. Open circuits file **d07-02b**. This circuit is the same as the circuit of Step 2 except that the outputs won't change (a connection problem?). What is

   wrong? The problem is _____

   _____

   _____.

5. Open circuits file **d07-02c**. This circuit is the same as the circuit of Step 2 except that one set of outputs (~1$Q$ and ~2$Q$) stays on all of the

   time (a circuit problem?). What is wrong? The problem is _____

   _____

   _____.

● *Circuit Construction:*

6. Open circuits file **d07-02d**. Using the components on the workspace, construct a circuit similar to Figure 7-3. What is the frequency of the CLOCK signal? The frequency of the CLOCK signal is

   _____ kHz.

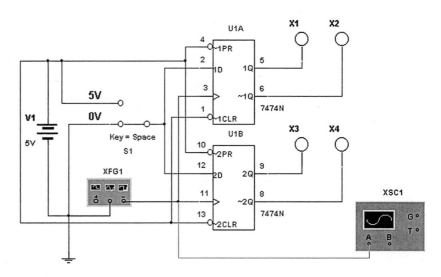

**Figure 7-3**   Constructed D Flip-Flop Circuit Using 7474N ICs

## Activity 7.3: The Edge-Triggered J-K Flip-Flop

1. The **Edge-Triggered J-K Flip-Flop** circuit is one of the most widely used type of flip-flop IC in the electronics industry. The circuit, in operation, is similar to the J-K Master-Slave Latch circuit, the only major difference being edge-triggering with a CLOCK input rather than (voltage) level-sensitivity enabling by means of a logic level input. Figure 7-4 displays the pinouts of the two sections of a 74LS7112N Dual J-K Flip-Flop.

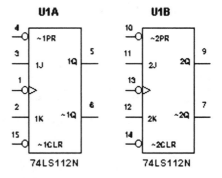

**Figure 7-4**   74LS112N Dual J-K Flip-Flop

2. Open circuits file **d07-03a**. In this circuit, the action of the 74LS112N flip-flop is examined. The 74LS112N is a true edge-triggered flip-flop rather than level actuated like the 7476. Actuate the switches and enter the resultant data in Table 7-2. When the switches are open, a HIGH is being applied to the associated input pin.

| 74LS112N J-K Flip-Flop Truth Table | | | | | | | |
|---|---|---|---|---|---|---|---|
| Inputs | | | | | Outputs | | |
| $\overline{PR}$ | $\overline{CLR}$ | CLK | J | K | Q | $\overline{Q}$ | Comments |
| 0 | 1 | X | X | X | | | RESET |
| 1 | 0 | X | X | X | | | SET |
| 0 | 0 | X | X | X | X | x | No Change |
| 1 | 1 | ↓ | 0 | 0 | X | x | No Change |
| 1 | 1 | ↓ | 1 | 0 | | | Q |
| 1 | 1 | ↓ | 0 | 1 | | | $\overline{Q}$ |
| 1 | 1 | ↓ | 1 | 1 | X | X | Toggle |

**Note:** ↓ indicates the trailing edge of the clock pulse, the positive to negative transition that causes the flip-flop to change states; X indicates that it doesn't matter what happens; x indicates that something will happen that is not totally predictable (the output might or might not change). The term toggle means that the flip-flop turns on and then off (goes HIGH and LOW) with the CLOCK signal.

**Table 7-2**  74LS112N Dual J-K Flip-Flop Truth Table

3. Open circuits file **d07-03b**. Using this circuit, we are going to look at the propagation delay of the 74LS112N IC. Perform the procedure as follows. This is a difficult project and will tax your abilities to use the oscilloscope. If you have problems, repeat the procedure until you get good results.

   a. Start by making sure that switch J1 (Spacebar) is closed (down) and switch J2 (C) is open (up).

   b. Activate the circuit and close switch C.

   c. Bring up the scope (double-click on the scope face) and then click back on the workspace.

   d. At this point, you should see a blue trace (square wave) in the upper half of the scope face and a red trace (a straight line) just below the blue trace.

   e. Place the mouse pointer on the pause button to the left of the on-off switch (upper right).

   f. As the trace starts at the left of the screen, switch J1 (spacebar) to the upper position.

   g. When the lower (red) trace goes down, click on the pause button with the mouse.

   h. At this point, you should have a display similar to Figure 7-5. Keep trying until you do.

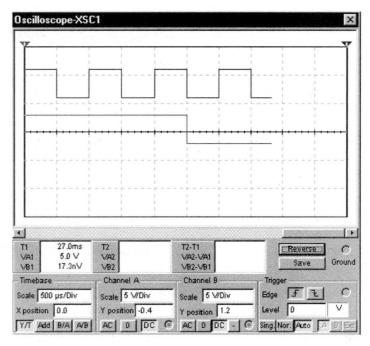

**Figure 7-5**  Oscilloscope Display of Flip-Flop
and Timing Signals

i.  Notice that the Timebase Scale is set at 500 μs/Div. Click in the Timebase Scale window and an "up/down" bar should appear similar to Figure 7-6.

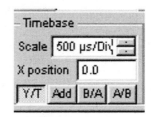

**Figure 7-6**  Timebase/Scale Up/Down Bar on Oscilloscope

j.  Click on the down arrow with your mouse and change the scale from 500 μsec to 200 μsec. You are now interested in the red trace and the falling portion of the waveform, which should have moved to the right. If the falling edge remains on the scope face, click on the down button again (the step down is 100 μsec).

k.  When the portion of the trace that we are interested in moves off the scope to the right, slowly click on the scroll bar arrow at the bottom of the scope until the falling (trailing) edge returns to the display. If the red trace goes below the middle horizontal line, you have gone too far. If that happens, retrace your steps using the up button in the Timebase Scale window. Figure 7-7 displays the scroll bar and arrow.

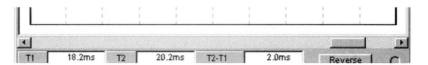

**Figure 7-7**    Oscilloscope Scroll Bar

l.  Keep clicking on the Scale button and the scroll bar until you have a display similar to Figure 7-8.

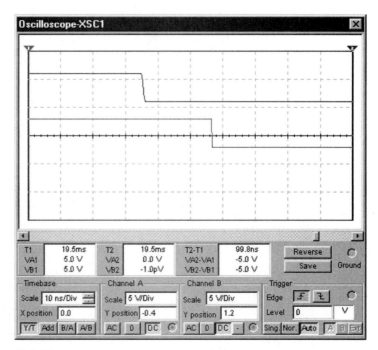

**Figure 7-8**    An Oscilloscope Display of Propagation Delay

m.  At this point, you are ready to make a propagation delay time measurement. Use the red and blue cursors to the left and right of the screen, and measure the time between the falling edge of the red and blue signals (if the cursor is yellow and to the left of the screen place the mouse on it and move it).

4.  What is the propagation delay between the input CLOCK signal and the change in the output signal of the flip-flop? The propagation delay is

about _____ nanoseconds.

### ● *Troubleshooting Problems:*

5. Open circuits file **d07-03c**. This 4027BP Dual J-K Flip-Flop CMOS circuit should be going HIGH and LOW (toggling), but it is hung up.

   What is the problem? The problem is that _____

   _____

   _____.

6. Open circuits file **d07-03d**. This 74LS112N circuit should be going HIGH and LOW (toggling), but it is hung up. What is wrong? The

   problem is that _____

   _____

   _____.

## Activity 7.4: Toggle (Trigger) Flip-Flops

1. The **Toggle** action of a flip-flop refers to the ability of a flip-flop to change states in sequence with the leading or trailing edge of an input clock signal. When toggle action is desired with an edge-triggered J-K flip-flop, the J-K inputs are rendered ineffective by connecting them HIGH (or LOW, depending on the characteristics of the IC). Also, the clear and preset inputs have to be rendered ineffective by tying them to 0 V or +5 V as necessary. When all of this is done, the $Q$ output follows or changes state with the respective leading or trailing edge of the input clock signal. At that point, the output signal will be one half of the frequency of the input clock signal. This is a two-to-one step down in frequency. Another term used for the toggle category of flip-flops is **Trigger** flip-flop.

2. Open circuits file **d07-04a**. The 7474N Dual D Master-Slave flip-flop circuit is one of the simplest to use in the toggle mode. The $\overline{Q}$ output is connected back to the D input, the $\overline{CLR}$ and $\overline{PR}$ inputs are connected HIGH, and the CLK input is connected to the input (trigger) signal source. Output $Q$ will toggle HIGH and LOW at one half of the frequency of the input signal source. Use the oscilloscope to determine the output frequency of the flip-flop circuit. What is that frequency? The

   frequency of the flip-flop circuit is _____ Hz. What is the frequency of the signal source? The frequency of the signal source is

   _____ Hz. Remember, the formula for the frequency of a waveform is Frequency = 1/Period (time of a complete square wave).

3. Open circuits file **d07-04b**. This 4027BD CMOS Dual J-K flip-flop circuit is constructed in the toggle mode. Notice that the J-K inputs are

connected HIGH and the CD and SD inputs are connected LOW. These connections allow the input clock signal to toggle the flip-flop. What is the output frequency of the flip-flop? The output frequency is

_____ Hz. What is the input frequency of the signal source? The

frequency of the input signal source is _____ Hz.

### ● *Troubleshooting Problems:*

4. Open circuits file **d07-04c**. This 4027BD CMOS Dual J-K flip-flop circuit has a problem, the outputs do not change. What is wrong? The

   problem is _____

   _____

   _____.

5. Open circuits file **d07-04d**. This 7474N Dual J-K flip-flop circuit has a problem. The $Q$ output does not change. What is wrong? The problem is

   _____

   _____.

### ● *Advanced Application:*

6. Open circuits file **d07-04e**. This circuit uses three 74LS107N integrated circuits to divide the frequency of the input signal by eight. Answer the following questions:

   a. What is the frequency of the input signal? The input frequency is

      _____ kHz. The measurement of the duration of one input CLOCK pulse is demonstrated in Figure 7-9. The duration of the

      CLOCK signal in this circuit is approximately _____ μseconds (trailing edge to trailing edge).

   b. To measure pulse duration (time), it is necessary to place the red and blue cursors at the trailing edges of two adjacent square waves to measure the amount of time between them. This time relationship is displayed in the T2-T1 window below the screen on the scope (in this example, 202.7 μseconds).

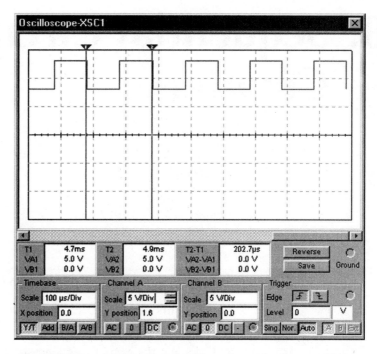

**Figure 7-9**    Measuring the Duration of a Square Wave

c.  What is the frequency of the output signal of IC U1A? The output frequency of flip-flop circuit U1A is _____ kHz. The duration of the output square wave for U1A is _____ μseconds.

d.  What is the frequency of the output signal of IC U1B? The output frequency of flip-flop circuit U1B is _____ kHz. The duration of the output square wave for U1B is _____ μseconds.

e.  What is the frequency of the output signal of IC2A? The output frequency of flip-flop circuit U3A is _____ kHz. The duration of the output square wave for U3A is _____ milliseconds.

f.  Does the 74LS107N use the leading or the trailing of the input pulse for triggering purposes? The 74LS107N uses the _____ edge of the input pulse.

7.  Obviously, the circuit met its challenge and the frequency of the input square wave was divided by eight. The input frequency of _____ Hz was changed to an output frequency of _____ Hz.

# 8. Sequential Circuits: Counters

**References**
*MultiSIM* 2001
*MultiSIM* 2001 Study Guide

**Objectives**   After completing this chapter, you should be able to:

- Determine the parameters of an asynchronous counter.
- Ascertain the modulus of a counter.
- Calculate the output frequency of a modulo (MOD) counter given the input clock frequency.
- Determine the count sequence, timing diagram and modulus of any counter.
- Manipulate up-down counters.
- Use IC counters as dividers (MOD counters).
- Use presettable counters.
- Use Johnson (ring) counters.
- Troubleshoot counters.

## Introduction

**Counter** circuits essentially count pulse inputs using various combinations of internal flip-flops to do so. The pulse inputs are the results of other types of digital circuit functions or external inputs from other sources such as sensors. An external input, for example, might be cases of peaches proceeding down an assembly line into a warehousing facility and being counted for inventory purposes. Counters and related circuits may be used to count items, to time functions, to synchronize various events, to divide, and to control results based upon the outcome of a counting function. The different types of counters are usually specified by their activity and the type of output they provide.

Counters can be further categorized as **Up** or **Down** counters. Up counters ascend from a count of zero or from a preset number, and down counters descend from a preset number or a maximum count. All of the counters that are discussed in this study will be categorized as up counters unless otherwise stated.

There are many types of counter circuits in digital equipment. We will be studying many of these circuits in this chapter. Among those we study will be:

- Asynchronous (ripple) counters

- Synchronous counters

- Presettable counters

- Up/Down counters

- Modulo (MOD) counters (asynchronous and synchronous)

- Johnson or ring counters

## Activity 8.1: The Asynchronous (Ripple) Counter

1. The **Asynchronous** (ripple) counter circuit uses a synchronous clock input for the first counter stage and then uses the $Q$ output of each of the following flip-flops in the counter as the input to the next flip-flop in the series. The counter stages following the first stage are classified as asynchronous; in other words, they are not exactly synchronized with the applied clock pulses. A basic four-stage ripple counter is shown in Figure 8-1. In this figure, the output frequency would be equal to the input frequency divided by 16. Each stage divides the input frequency by 2; thus an input frequency of 16 kHz would result in an output frequency of 1 kHz (for the circuit of Figure 8-1). The clock signal causes changes to occur in a ripple effect through the various stages of the counter in an asynchronous manner.

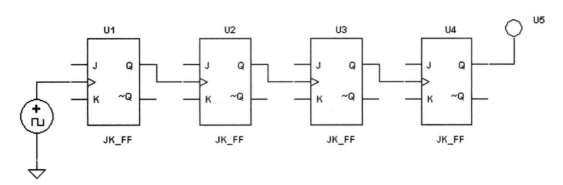

**Figure 8-1**   A Basic Asynchronous Counter

2. The counter of Figure 8-1 has sixteen distinctly different states, 0000 through 1111, thus it is called a **MOD-16 (modulo)** ripple counter. The **modulus** of a counter is defined as the number of states through which a counter sequences before repeating. The modulus is always equal to the number of possible states that it goes through before being returned to its starting condition. The overall modulus can be made larger or smaller by adding or subtracting flip-flops from the existing counter configuration. The modulus can be ascertained by counting the number, in input pulses, required to produce one output pulse or by dividing the input frequency

by the output frequency. The output frequency can be determined by measuring the duration (time) of one cycle of the pulse.

3. You can use the red and blue cursors on the oscilloscope display to measure the duration of any displayed signal. Using the mouse, place Cursor 1 (red) on the leading edge (where the waveform goes up) of the waveform and Cursor 2 (blue) on the leading edge of the next waveform. The leading-edge technique is displayed in Figure 8-2. The duration of the square wave is displayed in the third window under the screen (T2-T1) and, in this example, reads 175.2 μs (in this example, the frequency [F = 1/T] is 5707.8 Hz). An easy method to determine the modulo of a counter is to add up the binary count of the stages used to provide the ANDed reset pulse. For instance a four-stage counter, where stages 2 and 4 are being used to reset the counter, would be a MOD-6 counter.

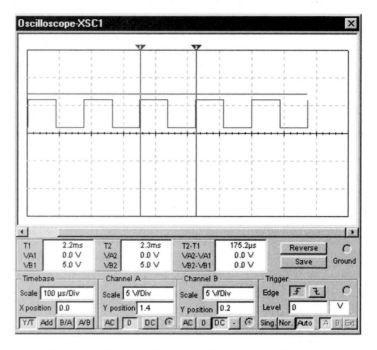

**Figure 8-2** Oscilloscope Display to Measure
Pulse Duration (Time)

4. Open circuits file **d08-01a**. In this three-stage asynchronous counter circuit, the input frequency is divided by _____. The input frequency of _____ kHz would result in an output frequency of _____ kHz. One cycle of the output waveform is _____ msec in duration. This is a MOD _____ counter.

5. Activate the circuit and draw the input and output waveforms displayed on the oscilloscope on Figure 8-3. Label the waveforms.

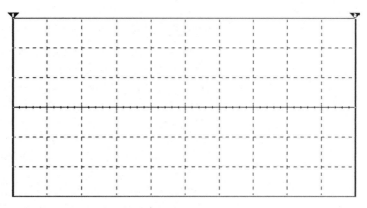

**Figure 8-3** Oscilloscope Display for an
Asynchronous Counter

6. Open circuits file **d08-01b**. This circuit uses a two-stage J-K flip-flop IC

   as an asynchronous counter. The circuit is a MOD _____ ripple
   counter. What are the input and output frequencies. The input frequency

   is _____ kHz and the output frequency is _____ kHz (the output
   frequency is equal to the input frequency divided by the modulus of the
   circuit).

7. Use the cursor technique to measure the duration of the input square
   wave of circuit **d08-01b**. The duration of the output square wave of

   this circuit is _____ msec. Calculate the frequency. The calculated fre-

   quency is _____ kHz. Is this frequency the same as the frequency of

   the circuit of Step 5? Yes/No _____. The answer should be yes. Explain

   the reason for a yes answer:_____

   _____

   _____.

8. The 74LS293 is a four-stage asynchronous counter integrated circuit.
   The four stages provide a MOD-16 (divide by 16) output or with the
   proper gating, can provide a modulus (divide by function) for any number less than 17. The 74LS293 is displayed in Figure 8-4. The outputs are
   QA, QB, QC, and QD (pins 3, 4, 5, and 9). The reset inputs R0(1) and
   R0(2) both have to be high (+5 V) simultaneously for the counter to be
   reset to the zero state (0 V). Clock input A (an active LOW) clocks the
   stage 1 of the counter and clock input B (an active LOW) clocks stages
   2 through 4 of the counter (asynchronously). If the counter is to be used

as a four-stage asynchronous counter, it is necessary to connect the output of stage 1 (Q0–pin 3) to the clock input of stage 2 (B–pin 11).

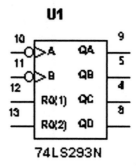

**Figure 8-4**   The 74LS293 (Four Stage) Asynchronous Counter

9. Open circuits file **d08-01c**. This circuit uses the 74LS293 as a MOD-14 counter. How many pulses does the circuit count before it returns to the

   zero state? This counter circuit counts _____ pulses before it resets to the zero state. What is the frequency of the output square wave? The out-

   put frequency is _____ kHz.

10. Binary counters can be used to divide an input frequency into subfrequencies that can be used elsewhere in digital equipment.

11. Open circuits file **d08-01d**. This circuit uses a twelve-stage asynchronous counter to develop a 1000 Hz signal from a 120 kHz signal. In other words, this is a divide-by-120 circuit. Use the cursors to measure and record the duration of the output pulse with the oscilloscope. The

    duration of the output pulse is _____ msec.

12. Change the frequency of this input clock signal to 240 kHz. Measure and record the duration of the output pulse. The duration of the output pulse

    is _____ μsec. Does the counter divide by 120 when you change the

    input clock frequency? Yes/No _____.

## ● *Troubleshooting Problems:*

13. Open circuits file **d08-01e**. The intent of this circuit is to use the 74LS293 as a MOD-12 counter. Check the modulus with the oscillo-

    scope. The counter is counting _____ pulses before it returns to the

    zero state. It is acting as a MOD _____ counter. What is wrong?

(*Note:* this is a connection problem, not a circuit fault.) The wire connected to Pin _____ on the 74LS293N should be connected to Pin _____. The output frequency is _____ Hz instead of _____ Hz.

14. Correct the problem. Now the counter is a MOD _____ counter and is counting _____ pulses before it resets to the zero state. Do not save the modified circuit when the screen prompts you for a decision.

15. Open circuits file **d08-01f**. This circuit is using the 74LS293 and is supposed to be a MOD-11 counter. Check the modulus with the oscilloscope. The counter is _____.
What is wrong? (*Note:* this is a circuit fault rather than a connection problem.) The problem is _____

_____

_____.

16. Open circuits file **d08-01g**. This circuit uses two 74LS73N integrated circuits to create a MOD-16 counter. The problem is that the count never reaches 16, it resets at a count of 8. What is the problem (look at the connections)? The problem is that _____

_____

_____.

Try to correct the problem. What did you do? I _____

_____

_____.

### ● *Circuit Construction:*

17. Open circuits file **d08-01h**. Using the components on the work-space, construct a MOD-75 counter. Draw your constructed circuit on Figure 8-5.

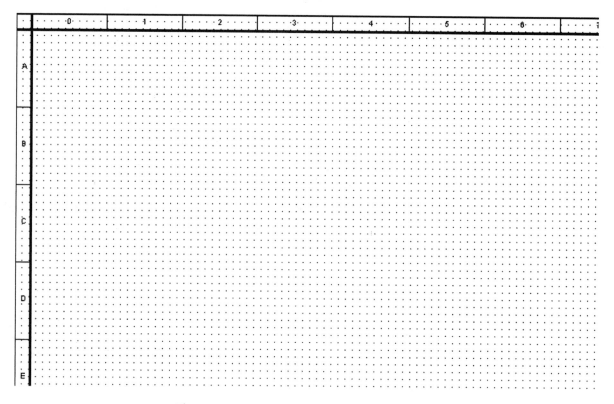

**Figure 8-5** Constructed MOD-75 Counter

## Activity 8.2: The Synchronous Counter

1. A **Synchronous** counter is a counter whose flip-flops are clocked simultaneously by a common clock source. The synchronization of the clocking event causes the transitions of all of the flip-flop states to occur simultaneously. A three-stage synchronous counter is displayed in Figure 8-6. Notice that the input timing signal is applied to the clock inputs on the J-K flip-flops and the J-K inputs are used for the counting function.

2. Open circuits file **d08-02a**. In this three-stage synchronous counter circuit the input frequency is divided by _____. The input frequency of _____ kHz would result in an output frequency of _____ kHz. One cycle of the output waveform is _____ msec in duration. This is a MOD _____ counter.

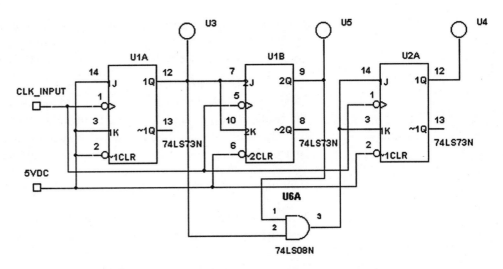

**Figure 8-6**   Three-Stage Synchronous Counter

3. Actuate the circuit and draw the input and output waveforms that you see displayed on the oscilloscope in Figure 8-7. Label the waveforms.

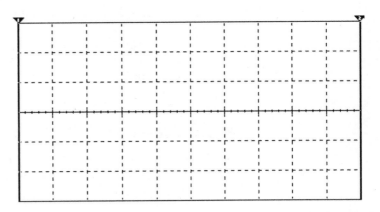

**Figure 8-7**   Oscilloscope Display for Synchronous Counter

4. Open circuits file **d08-02b**. This circuit uses one section of a 74LS393N Dual Binary Counter IC as a synchronous counter. The circuit is a MOD

   _____ counter. What are the input and output frequencies. The input

   frequency is _____ kHz and the output frequency is _____ Hz.

5. Use the cursors on the oscilloscope to determine the duration of one cycle of the output signal. The duration of one output cycle is

   _____ msec. Calculate the output frequency (F = 1/T). The calculated

   frequency is _____ Hz. Does this agree with the output frequency of

   Step 4? Yes/No _____.

6. When it is necessary to increase the modulus beyond the capabilities of an individual counter, additional counters can be added in series with the

first counter to multiply the modulus capabilities of the first counter. For example, a MOD-16 counter output can be multiplied by a MOD-2 (next counter stage) counter for an overall count of MOD-32.

7. Open circuits file **d08-02c**. This circuit uses two sections of a 74LS393N Dual Binary Counter IC to multiply the modulus of the first section by the modulus of the second section. To make the count an easy matter, we will use individual MOD-6 and MOD-3 sections for an overall modulus of 18. Measure the duration of the output square wave. The duration of

   the output square wave is _____ msec. What is the calculated output

   frequency? The calculated output frequency is _____ Hz. Count the number of input square waves for one cycle of the output square wave.

   This circuit is a MOD _____ counter.

8. Open circuits file **d08-02d**. This is a frequency divider circuit; the input signal is divided by the modulus of the overall counter circuit. Determine the frequency of the output signal. The frequency of the output signal is

   _____ Hz. What is the duration of the output signal? The duration

   of the output signal is _____ msec. This circuit is a MOD

   _____ counter.

### ● *Troubleshooting Problems:*

9. Open circuits file **d08-02e**. This circuit is similar to the circuit of Step 7. It is supposed to divide the input frequency by 100, but it isn't. Determine what the actual modulus of the circuit is and where the circuit

   connection is in error. The modulus of this circuit is _____. The

   connection problem with the circuit is _____

   _____

   _____.

10. Open circuits file **d08-02f**. This MOD-16 counter isn't working properly.

    What is wrong? The problem is that _____

    _____

    _____.

● *Circuit Construction:*

11. Open circuits file **d08-02g**. Using the components on the work-space, construct a MOD-90 counter. Draw your constructed circuit on Figure 8-8.

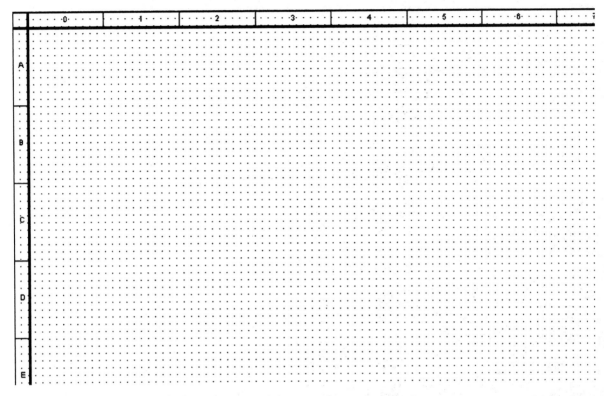

**Figure 8-8**   Circuit for a MOD-80 Counter

12. What is the frequency of the output signal? The frequency of the

    output signal is _____ kHz.

## Activity 8.3: Synchronous Down and Up/Down Counters

1. The count of a Down counter descends from a maximum or predeter-mined (preset) point determined by the modulus of the circuit or from the preset number. A down counter is simple to construct. It is simply a mat-ter of applying the $\overline{Q}$ outputs of all flip-flop stages to the J-K inputs of the next stage, rather than the $Q$ outputs. Then the count of the counter will descend instead of ascending. Many down counters will cycle from a count of zero to the maximum and then start counting down again from that maximum.

2. Open circuits file **d08-03a**. This is a simple two-stage down counter. Its count starts at zero, returns to the maximum count, and counts down to

zero again. Toggle the switch and enter the results in Table 8-1. What is the modulus of this circuit? The modulus of this circuit is _____.

| Switch Actuation | Output B | Output A | Decimal Count |
|---|---|---|---|
| 0 | 0 | 0 | 0 |
| 1 | | | 3 |
| 2 | | | 2 |
| 3 | | | 1 |
| 4 | | | 0 |

**Table 8-1**  Down Counter Data

3. Open circuits file **d08-03b**. This down counter, using the 74LS76N J-K Flip-Flop IC, is designed to count down from 12 to 3. At the detection of a count of 3, the counter resets back to 12 and starts counting down again. Toggle the input switch and record the LED outputs for each actuation of the switch. Record your data in Table 8-2. Notice that the initial condition (12) is determined by whether the CLR or PR pin is active (LOW). If the CLR or PR pin is not being used for a particular stage, it is connected to +5 VDC.

| Switch Actuation | LED 4 | LED 3 | LED 2 | LED 1 | Decimal Count |
|---|---|---|---|---|---|
| 0 | 1 | 1 | 0 | 0 | 12 |
| 1 | | | | | |
| 2 | | | | | |
| 3 | | | | | |
| 4 | | | | | |
| 5 | | | | | |
| 6 | | | | | |
| 7 | | | | | |
| 8 | | | | | |
| 9 | | | | | |

**Table 8-2**  Presettable Down Counter Data

4. Up/Down counters are usually found as four-stage ICs with the up or down function being selected externally. Typical ICs (74LS family) that operate as Up/Down binary counters are the 74LS161N, 74LS163N, 74LS191N, and the 74HC393N.

5. Open circuits file **d08-03c**. This circuit is built around the 74LS191N Synchronous Binary Up/Down 4-Bit Counter. It is designed to preload a count into the flip-flops and then allow the counter to count up or down from that point. This circuit is a complex circuit and requires study of the external connections to be able to fully understand its operation. The switches operate according to Table 8-3.

| **Switch D** | Data that will be placed into A, B, C, and D flip-flops<br>Open:  Loads all HIGHs into counter<br>Closed:  Loads all LOWs into counter |
|---|---|
| **Switch L** | Loads data when actuated—overrides existing counter data<br>Open:  Does not affect circuit<br>Closed:  Loads data into counter flip-flops according to data input |
| **Switch U** | Determines if counter will be up-count or down-count<br>Open:  Counter will be down-count<br>Closed:  Counter will be up-count |
| **Switch C** | Input clock pulse switch—toggles HIGH and LOW for each input |
| **Switch N** | Determines whether counter will count clock inputs<br>Open:  Will not count clock inputs<br>Closed:  Clock signal input position |

**Table 8-3**    74LS191N Up/Down Counter Circuit Switch Data

6. Actuate the circuit according to the procedure of Table 8-4 on page 108. Enter the resultant data into the table. Turn on the circuit first and then begin the procedure. D-U in the table means to toggle the switch down and then up. U-D means toggle up and then down.

● *Troubleshooting Problems:*

7. Open circuits file **d08-03d**. This is the same circuit as Steps 5 and 6. The circuit won't load HIGH data into the counter flip-flops. The counter works okay. What is wrong with the data load portion of the circuit? The

problem is that _____

_____

_____.

| Step | Switch Position | | | | | Data Indicators | | | |
|---|---|---|---|---|---|---|---|---|---|
| | D | L | U | C | N | A | B | C | D |
| Initial | Up | Up | Up | Up | Up | 0 | 0 | 0 | 0 |
| 1 | | D-U | | | | 1 | 1 | 1 | 1 |
| 2 | | | | | Down | | | | |
| 3 | | | | D-U | | | | | |
| 4 | | | | D-U | | | | | |
| 5 | | | | D-U | | | | | |
| 6 | | | | D-U | | | | | |
| 7 | | | | D-U | | | | | |
| 8 | | | | D-U | | | | | |
| 9 | | | | D-U | | | | | |
| 10 | | | | D-U | | | | | |
| 11 | | | | D-U | | | | | |
| 12 | | | | D-U | | | | | |
| 13 | | | | D-U | | | | | |
| 14 | | | | D-U | | | | | |
| 15 | | | | D-U | | | | | |
| 16 | | | | D-U | | | | | |
| 17 | | | | D-U | | | | | |
| 18 | | | | D-U | | | | | |
| 19 | | | | | Up | | | | |
| 20 | Down | | | | | 1 | 1 | 1 | 1 |
| 21 | | D-U | | | | 0 | 0 | 0 | 0 |
| 22 | | | Down | | | | | | |
| 23 | | | | | Down | | | | |
| 24 | | | | D-U | | | | | |
| 25 | | | | D-U | | | | | |
| 26 | | | | D-U | | | | | |
| 27 | | | | D-U | | | | | |
| 28 | | | | D-U | | | | | |
| 29 | | | | D-U | | | | | |
| 30 | | | | D-U | | | | | |
| 31 | | | | D-U | | | | | |
| 32 | | | | D-U | | | | | |
| 33 | | | | D-U | | | | | |
| 34 | | | | D-U | | | | | |
| 35 | | | | D-U | | | | | |
| 36 | | | | D-U | | | | | |
| 37 | | | | D-U | | | | | |
| 38 | | | | D-U | | | | | |
| 39 | | | | D-U | | | | | |
| 40 | Up | | Up | | Up | 0 | 0 | 0 | 0 |

**Table 8-4**   74LS191N   Up/Down Counter Circuit Operating Data

**Note:** D-U indicates a movement of the switch downward and then up.

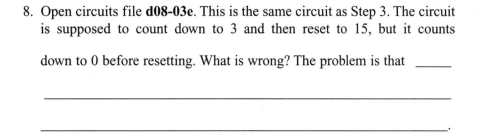

8. Open circuits file **d08-03e**. This is the same circuit as Step 3. The circuit is supposed to count down to 3 and then reset to 15, but it counts

down to 0 before resetting. What is wrong? The problem is that _____

_____

_____.

## ● *Circuit Construction:*

9. Open circuits file **d08-03f**. Using the components on the workspace, construct a down counter that will count from 7 to 2 and reset back to 0 and start over again. Draw your constructed circuit on Figure 8-9. This circuit is partially constructed.

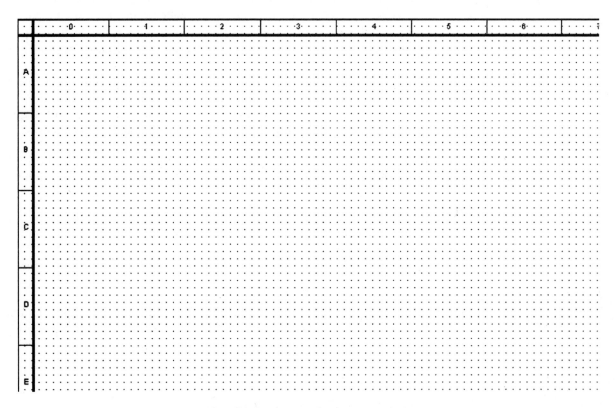

**Figure 8-9**   Circuit for Down Counter

# Activity 8.4: Presettable Counters

1. **Presettable** counters are used to load a predetermined count into a counter and to continue counting from that preset point. Most IC counters are presettable.

2. Open circuits file **d08-04a**. This presettable counter circuit is built around the 74LS163N, a Synchronous 4-Bit Binary Counter. It is designed to load a predetermined count into the counter flip-flops and then allow the counter to count up from that preset point. Switches A–D

determine the HIGH or LOW data that is to be loaded into the counter. The procedure to use the circuit is as follows:

- Turn on the circuit.

- Set the data switches (A through D) to the desired input data configuration.

- Close (down) the Load switch (L).

- Toggle (down and then up) the CLOCK switch (spacebar).

- Open the load switch (up).

- Toggle the CLOCK switch and watch the counter count up from the preset count.

3. Operate the circuit until it works according to the procedure of Step 2. Once you understand how the circuit works, load A and B HIGH (both switches up) and C and D LOW (both switches down). Now toggle the clock switch (down and then up) three times. What is the decimal equivalent to the count in the counter at this point? The decimal equivalent is

_____. Now toggle the clock switch ten more times. What is the decimal equivalent now? The decimal equivalent is _____. If you toggle the clock switch four more times the decimal count should be 4.

4. Open circuits file **d08-04b**. This circuit uses four 74HC76N flip-flops and other ICs to create a four-stage presettable counter. The counter switches operate as follows:

- Switches A, B, C, and D are the data switches—they determine the preset count to be loaded into the counter.

- Switch L is the load switch, it has priority over the CLOCK switch.

- Switch T is the CLOCK switch.

5. Activate the circuit switches according to Table 8-5 on page 111 and enter the resultant data into the table. When you change switch positions, start at the left and progress to the right according to the table.

## ● *Troubleshooting Problems:*

6. Open circuits file **d08-04c**. This is the same circuit as Step 2 except that the circuit won't load or count. What is wrong (this is a circuit fault)? The problem is _____

_____

_____.

| Step | Switch Positions | | | | | | Data Indicators | | | |
|---|---|---|---|---|---|---|---|---|---|---|
| | A | B | C | D | L | T | A | B | C | D |
| Initial | D | D | D | D | D | U | 0 | 0 | 0 | 0 |
| 1 | | | | | U-D | | | | | |
| 2 | U | | | | U-D | | | | | |
| 3 | D | U | | | U-D | | | | | |
| 4 | U | U | | | U-D | | | | | |
| 5 | D | D | U | | U-D | | | | | |
| 6 | U | | | | U-D | | | | | |
| 7 | D | U | | | U-D | | | | | |
| 8 | U | | | | U-D | | | | | |
| 9 | D | D | D | U | U-D | | | | | |
| 1 | 1 | | | | U-D | | | | | |
| 0 | D | U | | | U-D | | | | | |
| 11 | U | | | | U-D | | | | | |
| 12 | D | D | U | | U-D | | | | | |
| 13 | U | | | | U-D | | | | | |
| 14 | D | U | | | U-D | | | | | |
| 15 | U | | | | U-D | | | | | |
| 16 | D | D | D | D | U-D | | | | | |
| 17 | | | | | D-U | | | | | |
| 18 | | | | | D-U | | | | | |
| 19 | | | | | D-U | | | | | |
| 20 | | | | | D-U | | | | | |
| 21 | | | | | D-U | | | | | |
| 22 | | | | | D-U | | | | | |
| 23 | | | | | D-U | | | | | |
| 24 | | | | | D-U | | | | | |
| 25 | | | | | D-U | | | | | |
| 26 | | | | | D-U | | | | | |
| 27 | | | | | D-U | | | | | |
| 28 | | | | | D-U | | | | | |
| 29 | | | | | D-U | | | | | |
| 30 | | | | | D-U | | | | | |
| 31 | | | | | D-U | | | | | |
| 32 | | | | | D-U | | 0 | 0 | 0 | 0 |

**Table 8-5**   Presettable Counter Data

7. Open circuits file **d08-04d**. This is the same circuit as Step 2 except that the circuit won't load or count. What is wrong (this is a connection problem)? The problem is _____

_____

_____.


● *Circuit Construction:*

8. Open circuits file **d08-04e**. Use the components on the workspace to construct a presettable counter that will start at a count of 3 and count up to 12 before returning to the preset of 3.

# 9. Sequential Circuits: Shift Registers

**References**
*MultiSIM* 2001
*MultiSIM* 2001 Study Guide

**Objectives**   After completing this chapter, you should be able to:

- Construct shift register circuits using flip-flops.
- Distinguish between the four possible shift register combinations of:
  - Serial In/Serial Out
  - Serial In/Parallel Out
  - Parallel In/Serial Out
  - Parallel In/Parallel Out
- Use IC shift registers.
- Troubleshoot shift registers.

## Introduction

**Shift registers** circuits are synchronous digital circuits used to store or move binary data. These types of circuits consist either of a series of flip-flops in groups which can store one bit of data each or as shift register ICs that store larger groups of data. The amount of data that can be stored in a shift register depends on the width of the register and the number of storage flip-flops contained in the register. A typical shift register may contain four or eight stages with the inputs and outputs of data being controlled by the leading or trailing edge of a clock pulse. This synchronized movement of digital data in and out of shift register circuits can be accomplished by four basic methods:

- Serial In/Serial Out, where the data is placed in the register in a bit-by-bit method over a single conductor and taken out of the register in the same bit-by-bit method as shown in Figure 9-1.

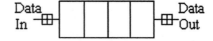

**Figure 9-1**   Serial In/Serial Out Shift Register

- Serial In/Parallel Out, where the data is placed in the register in a bit-by-bit method over a single conductor and taken out of the register from each flip-flop simultaneously in a parallel fashion as shown in Figure 9-2.

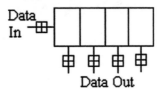

**Figure 9-2**   Serial In/Parallell Out Shift Register

- Parallel In/Serial Out, where the data is placed in the register flip-flops simultaneously in a parallel fashion, and the data is taken from the register in a bit-by-bit method over a single conductor as shown in Figure 9-3.

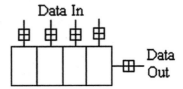

**Figure 9-3**   Parallel In/Serial Out Shift Register

- Parallel In/Parallel Out, where the data is placed in the register flip-flops simultaneously in a parallel fashion and taken out of the register by the same method as shown in Figure 9-4.

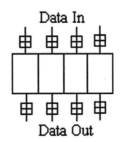

**Figure 9-4**   Parallel In/Parallel Out Shift Register

Shift registers are further categorized by their ability to shift data from left to right or from right to left. A shift register with this left-right, right-left shifting characteristic is referred to as having bidirectional capabilities.

Many of the projects in this chapter use the 74HC194 4-Bit Bidirectional Universal Shift Register. This shift register incorporates many of the shift register features that a circuit designer might desire. The register has four distinct modes of operation:

- Parallel Load

- Shift Left (with a serial load [SR])

- Shift Right (with a serial load [SL])
- Inhibit Clock (do nothing)

These modes of operation are selected with two distinct mode input terminals (S0 and S1) that key the desired mode using HIGH and LOW inputs as follows:

- S0 and S1 LOW inhibits clocking of the input (nothing happens)
- S0 LOW and S1 HIGH causes data to shift left (synchronously)
- S0 HIGH and S1 LOW causes data to shift right (synchronously)
- S0 and S1 HIGH causes data to load in parallel (A, B, C and D)

S0 and S1 have to be connected either HIGH or LOW, depending on the desired operation, or no data will be entered into the register.

## Activity 9.1: Serial In/Serial Out Shift Registers

1.  Serial In/Serial Out shift registers can be constructed using D flip-flops as displayed in Figure 9-5.

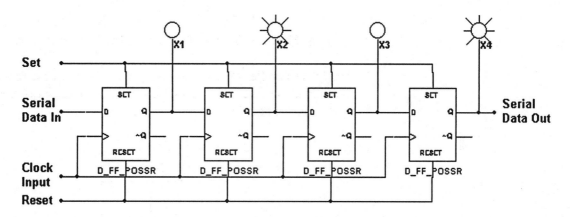

**Figure 9-5**   Serial In/Serial Out Shift Register Using D Flip-Flops

2.  Open circuits file **d09-01a**. In this circuit, S1 (C) is the Clear switch, S2 (spacebar) is the Clock input, and S3 (D) is the Data switch. Activate the switches according to Table 9-1 on page 116 and record the results. This procedure should have progressively turned on all four lamps one at a time as the Clock pulse moves the data into the register.

| Step | Switch Settings | | | Register Data Condition | | | |
|---|---|---|---|---|---|---|---|
| | **S1**<br>**Clear** | **S2**<br>**Clock** | **S3**<br>**Data** | **X1** | **X2** | **X3** | **X4** |
| Initial | HIGH | LOW | LOW | X | X | X | X |
| 1 | LOW | | | | | | |
| 2 | HIGH | | HIGH | | | | |
| 3 | | Toggle | | | | | |
| 4 | | Toggle | | | | | |
| 5 | | Toggle | | | | | |
| 6 | | Toggle | | | | | |

**Table 9-1**   Serial In/Serial Out Shift Register Data

3. Using the same circuits file, return the switches to the initial conditions of Table 9-2. Activate the switches according to the table and record the resultant data.

| Step | Switch Settings | | | Register Data Condition | | | |
|---|---|---|---|---|---|---|---|
| | **S1**<br>**Clear** | **S2**<br>**Clock** | **S3**<br>**Data** | **X1** | **X2** | **X3** | **X4** |
| Initial | HIGH | LOW | LOW | X | X | X | X |
| 1 | LOW | | | | | | |
| 2 | HIGH | | HIGH | | | | |
| 3 | | Toggle | | | | | |
| 4 | | Toggle | | | | | |
| 5 | | Toggle | | | | | |
| 6 | | Toggle | | | | | |
| 7 | | | LOW | | | | |
| 8 | | Toggle | | | | | |
| 9 | | Toggle | | | | | |
| 10 | | Toggle | | | | | |

**Table 9-2**   Additional Serial In/Serial Out Shift Register Data

*Table continues next page*

| Step | Switch Settings | | | Register Data Condition | | | |
|------|------------------|----------------|--------------|----|----|----|----|
|      | **S1**<br>**Clear** | **S2**<br>**Clock** | **S3**<br>**Data** | **X1** | **X2** | **X3** | **X4** |
| 11   |      | Toggle |      |    |    |    |    |
| 12   | LOW  |        |      |    |    |    |    |
| 13   | HIGH |        | HIGH |    |    |    |    |
| 14   |      | Toggle |      |    |    |    |    |
| 15   |      |        | LOW  |    |    |    |    |
| 16   |      | Toggle |      |    |    |    |    |
| 17   |      | Toggle |      |    |    |    |    |
| 18   |      | Toggle |      |    |    |    |    |
| 19   |      | Toggle |      |    |    |    |    |

**Table 9-2**   *(continued)*

4. After completing Table 9-2 try placing alternate data in the register, a HIGH, a LOW, a HIGH, and a LOW.

5. Open circuits file **d09-01b**. This circuit is similar to the circuit of Step 2, but uses the 74HC194N shift register (Figure 9-6) instead of individual flip-flops. Activate the circuit according to Table 9-3 and enter your results in the table.

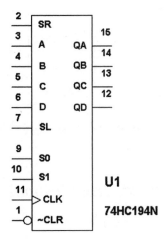

**Figure 9-6**   The 74HC194N Bidirectional Universal Shift Register

| Step | Switch Settings | | | Register Data Condition | | | |
|---|---|---|---|---|---|---|---|
| | S1 Clear | S2 Clock | S3 Data | X1 | X2 | X3 | X4 |
| Initial | HIGH | LOW | LOW | X | X | X | X |
| 1 | LOW | | | | | | |
| 2 | HIGH | | HIGH | | | | |
| 3 | | Toggle | | | | | |
| 4 | | Toggle | | | | | |
| 5 | | Toggle | | | | | |
| 6 | | Toggle | | | | | |
| 7 | | | LOW | | | | |
| 8 | | Toggle | | | | | |
| 9 | | Toggle | | | | | |
| 10 | | | | | | | |
| 11 | | Toggle | | | | | |
| 12 | LOW | | | | | | |
| 13 | HIGH | | HIGH | | | | |
| 14 | | Toggle | | | | | |
| 15 | | | LOW | | | | |
| 16 | | Toggle | | | | | |
| 17 | | Toggle | | | | | |
| 18 | | Toggle | | | | | |
| 19 | | Toggle | | | | | |

**Table 9-3**   Additional Serial In/Serial Out Shift Register Data

## ● *Troubleshooting Problems:*

6. Open circuits file **d09-01c**. This circuit has a connection problem, what is wrong? The problem with the circuit is _____

_____

_____.

7. Open circuits file **d09-01d**. This circuit has a circuit problem, what is wrong? The problem with the circuit is _____

_____

_____.

● *Circuit Construction:*

8. Open circuits file **d09-01e**. Use the components on the workspace to construct a serial in/serial out shift register similar to the circuit of Step 2.

## Activity 9.2:  Serial In/Parallel Out Shift Registers

1. Serial In/Parallel Out shift registers can be constructed using D flip-flops as displayed in Figure 9-7.

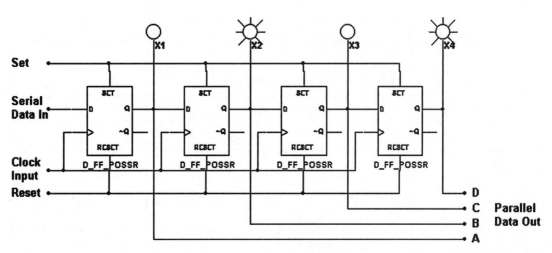

**Figure 9-7**    Serial In/Parallel Out Shift Register Using D Flip-Flops

2. Open circuits file **d09-02a**. This circuit uses 74LS74N ICs to construct a serial in/parallel out shift register. S1 (C) is the Clear switch, S2 (space-bar) is the Clock input, and switch 3 (D) is the Data switch. Activate the switches according to Table 9-4 on page 120 and record the results. This procedure should progressively turn on all four lamps one at a time. The outputs would be taken from the $Q$ outputs at the proper time as determined by the external circuits. The outputs continually reflect conditions of the register at any point in time.

| Step | Switch Settings | | | Register Data Condition | | | |
|---|---|---|---|---|---|---|---|
| | S1 Clear | S2 Clock | S3 Data | A | B | C | D |
| Initial | HIGH | LOW | LOW | X | X | X | X |
| 1 | LOW | | | | | | |
| 2 | HIGH | | HIGH | | | | |
| 3 | | Toggle | | | | | |
| 4 | | Toggle | | | | | |
| 5 | | Toggle | | | | | |
| 6 | | Toggle | | | | | |

**Table 9-4**   Serial In/Parallel Out Shift Register Data

3. Return the switches to the initial conditions of Table 9-5. Activate the switches according to the table and record the resultant data in the table.

| Step | Switch Settings | | | Register Data Condition | | | |
|---|---|---|---|---|---|---|---|
| | S1 Clear | S2 Clock | S3 Data | A | B | C | D |
| Initial | HIGH | LOW | LOW | X | X | X | X |
| 1 | LOW | | | | | | |
| 2 | HIGH | | HIGH | | | | |
| 3 | | Toggle | | | | | |
| 4 | | Toggle | | | | | |
| 5 | | Toggle | | | | | |
| 6 | | Toggle | | | | | |
| 7 | | | LOW | | | | |
| 8 | | Toggle | | | | | |
| 9 | | Toggle | | | | | |
| 10 | | Toggle | | | | | |
| 11 | | Toggle | | | | | |
| 12 | LOW | | | | | | |
| 13 | HIGH | | HIGH | | | | |

**Table 9-5**   Additional Serial In/Parallel Out Shift Register Data

*Table continues next page*

| Step | Switch Settings | | | Register Data Condition | | | |
|------|-----------------|---|---|-------------------------|---|---|---|
| | **S1** Clear | **S2** Clock | **S3** Data | **A** | **B** | **C** | **D** |
| 14 | | Toggle | | | | | |
| 15 | | | LOW | | | | |
| 16 | | Toggle | | | | | |
| 17 | | Toggle | | | | | |
| 18 | | Toggle | | | | | |
| 19 | | Toggle | | | | | |

**Table 9-5**    (*continued*)

4. Open circuits file **d09-02b**. This circuit is similar to the circuit of Step 2 but uses the 74HC194N shift register IC instead of individual flip-flop ICs. Activate the circuit according to Table 9-6 and enter the resultant data in the table.

| Step | Switch Settings | | | Register Data Condition | | | |
|------|-----------------|---|---|-------------------------|---|---|---|
| | **S1** Clear | **S2** Clock | **S3** Data | **X1** | **X2** | **X3** | **X4** |
| Initial | HIGH | LOW | LOW | X | X | X | X |
| 1 | LOW | | | | | | |
| 2 | HIGH | | HIGH | | | | |
| 3 | | Toggle | | | | | |
| 4 | | Toggle | | | | | |
| 5 | | Toggle | | | | | |
| 6 | | Toggle | | | | | |
| 7 | | | LOW | | | | |
| 8 | | Toggle | | | | | |
| 9 | | Toggle | | | | | |
| 10 | | | | | | | |
| 11 | | Toggle | | | | | |
| 12 | LOW | | | | | | |
| 13 | HIGH | | HIGH | | | | |

**Table 9-6**    Additional Serial In/Parallel Out Shift Register Data

*Table continues next page*

| Step | Switch Settings | | | Register Data Condition | | | |
|------|----------|----------|---------|----|----|----|----|
|      | **S1** **Clear** | **S2** **Clock** | **S3** **Data** | **X1** | **X2** | **X3** | **X4** |
| 14 |  | Toggle |  |  |  |  |  |
| 15 |  |  | LOW |  |  |  |  |
| 16 |  | Toggle |  |  |  |  |  |
| 17 |  | Toggle |  |  |  |  |  |
| 18 |  | Toggle |  |  |  |  |  |
| 19 |  | Toggle |  |  |  |  |  |

**Table 9-6**    (*continued*)

### ● *Troubleshooting Problems:*

5. Open circuits file **d09-02c**. This circuit has a circuit problem, what is wrong? The problem with the circuit is _____

_____

_____.

6. Open circuits file **d09-02d**. This 74HC194N circuit has a connection problem, what is wrong? The problem with the circuit is _____

_____

_____.

### ● *Circuit Construction:*

7. Open circuits file **d09-02e**. Use the components on the workspace to construct a serial in/serial out shift register similar to the circuit of Step 2.

## Activity 9.3: Parallel In/Serial Out Shift Registers

1. Parallel In/Serial Out shift registers can be constructed using D flip-flops as displayed in Figure 9-8 on page 123.

2. Open circuits file **d09-03a**. In this circuit, assembled with 74LS74N ICs, switch 1 (spacebar) is the Serial Shift Clock Input. Parallel data inputs are provided by switches D1 through D4, and switch 2 (P) determines the mode of the circuit (parallel in or serial out). When the Mode switch is

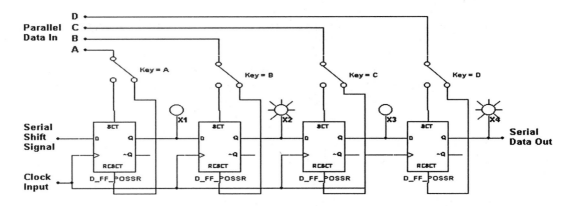

**Figure 9-8**   Parallel In/Serial Out Shift Register Circuit

in the up position, parallel inputs can be placed directly into the individual flip-flops; when the Mode switch is in the down position, data can be shifted to the right using the Serial Shift Clock Input. Activate the switches according to Table 9-7 and record the results.

| Step | Switch Settings | | | | | | Register Data Condition | | | |
|---|---|---|---|---|---|---|---|---|---|---|
| | S1 Serial Shift | S2 Mode | D1 | D2 | D3 | D4 | A | B | C | D |
| Initial | X | HIGH | 0 | 0 | 0 | 0 | 0 | 0 | 0 | 0 |
| 1 | X | | 1 | 1 | 1 | 1 | | | | |
| 2 | | LOW | | | | | | | | |
| 3 | Toggle | | | | | | | | | |
| 4 | Toggle | | | | | | | | | |
| 5 | Toggle | | | | | | | | | |
| 6 | Toggle | | | | | | | | | |

**Table 9-7**   Parallel In/Serial Out Shift Register Data

3. At this point, you should have loaded HIGHs into the register and then serially shifted the data out to the right, filling the register with LOWs. The reason that LOWs are entered into the register is that the data input of the first flip-flop is connected to 0 V. This LOW input could be connected to +6 V and HIGHs would be serially shifted into the register by the Serial Shift signal. Continue activating the switches according to Table 9-8 on page 124 and record your data.

| Step | Switch Settings | | | | | | Register Data Condition | | | |
|---|---|---|---|---|---|---|---|---|---|---|
| | S1 Serial Shift | S2 Mode | D1 | D2 | D3 | D4 | A | B | C | D |
| 7 | X | HIGH | | | | | 0 | 0 | 0 | 0 |
| 8 | X | | 1 | | | | | | | |
| 9 | X | LOW | | | | | | | | |
| 10 | Toggle | | | | | | | | | |
| 11 | Toggle | | | | | | | | | |
| 12 | Toggle | | | | | | | | | |
| 13 | Toggle | | | | | | | | | |

**Table 9-8**   More Parallel In/Serial Out Shift Register Data

4. Open circuits file **d09-03b**. This circuit uses the 74HC194N shift register IC to accomplish the parallel in/serial out task. Activate the circuit and then the switches according to Table 9-9 and record the resultant data.

| Step | Switch Settings | | | | | | | Register Data Condition | | | |
|---|---|---|---|---|---|---|---|---|---|---|---|
| | S1 Clear | S2 Clock | S3 In/Out | A | B | C | D | X1 | X2 | X3 | X4 |
| Initial | Up | Up | Open | 1 | 1 | 1 | 1 | X | X | X | X |
| 1 | | Toggle | | | | | | | | | |
| 2 | Down | | | | | | | | | | |
| 3 | Up | | | 1 | 0 | 1 | 0 | | | | |
| 4 | | Toggle | | | | | | | | | |
| 5 | | | Closed | | | | | | | | |
| 6 | | Toggle | | | | | | | | | |
| 7 | | Toggle | | | | | | | | | |
| 8 | | Toggle | | | | | | | | | |
| 9 | | Toggle | | | | | | | | | |
| 10 | | | Open | 1 | 0 | 0 | 0 | | | | |
| 11 | | Toggle | | | | | | | | | |
| 12 | | | Closed | | | | | | | | |

**Table 9-9**   Parallel In/Serial Out 74HC194N Shift Register IC Data

*Table continues next page*

| Step | Switch Settings | | | | | | | Register Data Condition | | | |
|------|------|------|------|---|---|---|---|----|----|----|----|
| | S1 Clear | S2 Clock | S3 In/Out | A | B | C | D | X1 | X2 | X3 | X4 |
| 13 | | Toggle | | | | | | | | | |
| 14 | | Toggle | | | | | | | | | |
| 15 | | Toggle | | | | | | | | | |
| 16 | | Toggle | | | | | | | | | |
| 17 | | | Open | 1 | 1 | 1 | 1 | | | | |
| 18 | | Toggle | | | | | | | | | |
| 19 | Down | | | | | | | | | | |

**Table 9-9**    (*continued*)

5. Notice that Step 1 places all of the flip-flops inside the register in the HIGH condition as displayed by the output indicators X1 through X4. Then Step 2 empties the register flip-flops with the Clear input overriding the register conditions.

6. The rest of the exercise consists of placing HIGHs and LOWs into the register and shifting them serially to the right.

7. Open circuits file **d09-03c**. In this circuit we are serially transferring the output of one register into another register. Activate the circuit and actuate the switches according to Table 9-10. Record the resultant data in the table.

| Step | Switch Settings | | | | | | | Register Data Condition | | | | | | | |
|------|------|------|------|---|---|---|---|------------|------|------|------|------------|------|------|------|
| | | | | | | | | Register 1 | | | | Register 2 | | | |
| | S1 Clear | S2 Clock | S3 In/Out | A | B | C | D | X1 | X2 | X3 | X4 | X5 | X6 | X7 | X8 |
| Initial | Up | Up | Open | 1 | 1 | 1 | 1 | X | X | X | X | X | X | X | X |
| 1 | | Toggle | | | | | | | | | | | | | |
| 2 | Down | | | | | | | | | | | | | | |
| 3 | Up | | | 1 | 0 | 0 | 0 | | | | | | | | |
| 4 | | Toggle | | | | | | | | | | | | | |
| 5 | | | Closed | | | | | | | | | | | | |
| 6 | | Toggle | | | | | | | | | | | | | |
| 7 | | Toggle | | | | | | | | | | | | | |

**Table 9-10**    Transferring Data from Register to Register

*Table continues next page*

| Step | Switch Settings | | | | | | | Register Data Condition | | | | | | | |
|------|------|------|------|---|---|---|---|---|---|---|---|---|---|---|---|
|  | | | | | | | | Register 1 | | | | Register 2 | | | |
|  | S1 Clear | S2 Clock | S3 In/Out | A | B | C | D | X 1 | X 2 | X 3 | X 4 | X 5 | X 6 | X 7 | X 8 |
| 8 |  | Toggle |  | | | | | | | | | | | | |
| 9 |  | Toggle |  | | | | | | | | | | | | |
| 10 |  | Toggle |  | | | | | | | | | | | | |
| 11 |  | Toggle |  | | | | | | | | | | | | |
| 12 |  | Toggle |  | | | | | | | | | | | | |
| 13 |  | Toggle |  | | | | | | | | | | | | |
| 14 |  |  | Open | | | | | | | | | | | | |
| 15 |  |  |  | 0 | 0 | 0 | 1 | | | | | | | | |
| 16 |  | Toggle |  | | | | | | | | | | | | |
| 17 |  | Toggle |  | | | | | | | | | | | | |
| 18 |  | Toggle |  | | | | | | | | | | | | |
| 19 |  | Toggle |  | | | | | | | | | | | | |
| 20 |  | Toggle |  | | | | | | | | | | | | |
| 21 | Down |  |  | | | | | | | | | | | | |

**Table 9-10**    *(continued)*

8. Notice that Step 4 enters a HIGH into Register 1, Flip-Flop A (a parallel transfer of data). Then Steps 5 through 13 serially transfer that HIGH through Register 1 (right shift), into the flip-flops of Register 2, and on out to the right.

9. The final steps demonstrate the serial shift-right of the HIGH condition of Register 1 Flip-Flop D into all of the flip-flops of Register 2. Each time the Clock input is toggled, Register 1 Flip-Flop D is set HIGH and then transferred to Register 2 the next time the Clock input is toggled.

## ● *Troubleshooting Problems:*

10. Open circuits file **d09-03d**. This circuit has a problem; the data will not serially shift out to the right. What is wrong? The problem with the

circuit is  _____

_____

_____ .

11. Open circuits file **d09-03e**. This parallel in, shift right, circuit has a connection problem, what is wrong? The problem with the circuit is

_____

_____

_____.

## ● *Circuit Construction:*

12. Open circuits file **d09-03f**. Use the components on the workspace to construct a parallel in/serial out shift register similar to the circuit of Figure 9-9. Then check its operation according to Step 4.

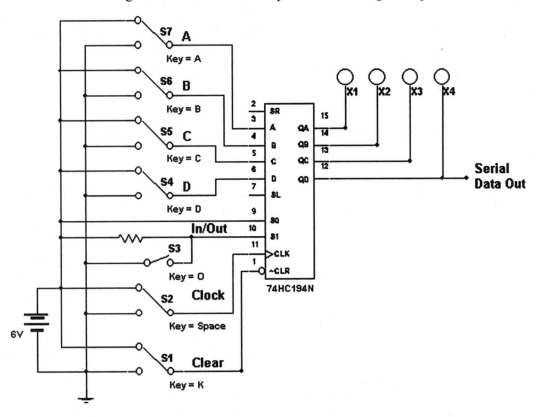

**Figure 9-9**   Construction Circuit for Parallel In/Serial Out Shift Register

## Activity 9.4: Parallel In/Parallel Out Shift Registers

1. Parallel In/Parallel Out shift registers can be constructed using D flip-flops as displayed in Figure 9-10 or by using the 74HC194N 4-Bit Bidirectional Universal Shift Register.

2. Open circuits file **d09-04a**. This circuit uses the 74HC194N Shift Register IC as a parallel in/parallel out shift register. Switch 1 (spacebar) is the Clock input and S2 (K) clears the register. Parallel data inputs are provided by switches J1 through J4 (A, B, C, and D); the mode of the circuit is determined by pins S0 and S1 of the IC which are wired to +5 V

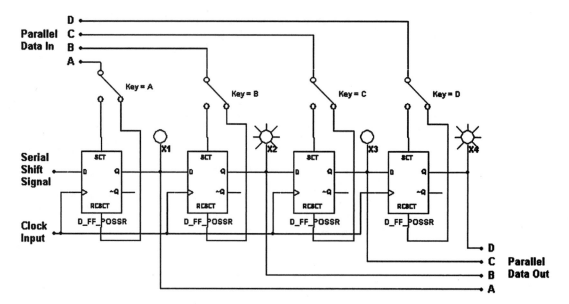

**Figure 9-10** Parallel In/Parallel Out Shift Register Circuit

(HIGH). The mode connections determine circuit operation as parallel in/parallel out. Activate the switches according to Table 9-11 and record the results.

| Step | | | Data Input | | | | Shift Register Output | | | |
|---|---|---|---|---|---|---|---|---|---|---|
| | **S2 Clear** | **S1 Clock** | **D** | **C** | **B** | **A** | **X1** | **X2** | **X3** | **X4** |
| Initial | Up | Down | 0 | 0 | 0 | 0 | | | | |
| | Toggle | | | | | | | | | |
| 1 | Up | Toggle | 0 | 0 | 0 | 1 | | | | |
| 2 | | Toggle | 0 | 0 | 1 | 0 | | | | |
| 3 | | Toggle | 0 | 0 | 1 | 1 | | | | |
| 4 | | Toggle | 0 | 1 | 0 | 0 | | | | |
| 5 | | Toggle | 0 | 1 | 0 | 1 | | | | |
| 6 | | Toggle | 0 | 1 | 1 | 0 | | | | |
| 7 | | Toggle | 0 | 1 | 1 | 1 | | | | |
| 8 | | Toggle | 1 | 0 | 0 | 0 | | | | |
| 9 | | Toggle | 1 | 0 | 0 | 1 | | | | |
| 10 | | Toggle | 1 | 0 | 1 | 0 | | | | |
| 11 | | Toggle | 1 | 0 | 1 | 1 | | | | |

**Table 9-11** Parallel In/Parallel Out Shift Register Data

*Table continues next page*

| Step | S2 Clear | S1 Clock | Data Input | | | | Shift Register Output | | | |
|---|---|---|---|---|---|---|---|---|---|---|
| | | | D | C | B | A | X1 | X2 | X3 | X4 |
| 12 | | Toggle | 1 | 1 | 0 | 0 | | | | |
| 13 | | Toggle | 1 | 1 | 0 | 1 | | | | |
| 14 | | Toggle | 1 | 1 | 1 | 0 | | | | |
| 15 | | Toggle | 1 | 1 | 1 | 1 | | | | |
| 16 | | Toggle | 0 | 0 | 0 | 0 | | | | |

**Table 9-11**   (*continued*)

3. Open circuits file **d09-04b**. Activate the circuit according to Table 9-12 and record the resultant data in the table.

| Step | Register Data Condition | | | | | | | | | | | | | |
|---|---|---|---|---|---|---|---|---|---|---|---|---|---|---|
| | Switch Settings | | | | | | Register 1 | | | | Register 2 | | | |
| | S1 Clear | S2 Clock | D | C | B | A | X1 | X2 | X3 | X4 | X5 | X6 | X7 | X8 |
| Initial | Up | Up | 1 | 1 | 1 | 1 | | | | | | | | |
| 1 | | Toggle | | | | | | | | | | | | |
| 2 | Down | | | | | | | | | | | | | |
| 3 | Up | | | | | | | | | | | | | |
| 4 | | | 0 | 0 | 0 | 1 | | | | | | | | |
| 5 | | Toggle | | | | | | | | | | | | |
| | | Toggle | | | | | | | | | | | | |
| 6 | | | 0 | 0 | 1 | 0 | | | | | | | | |
| 7 | | Toggle | | | | | | | | | | | | |
| 8 | | Toggle | | | | | | | | | | | | |
| 9 | | | 0 | 0 | 1 | 1 | | | | | | | | |
| 10 | | Toggle | | | | | | | | | | | | |
| 11 | | Toggle | | | | | | | | | | | | |
| 12 | | | 0 | 1 | 0 | 0 | | | | | | | | |
| 13 | | Toggle | | | | | | | | | | | | |
| 14 | | Toggle | | | | | | | | | | | | |
| 15 | | | 0 | 1 | 0 | 1 | | | | | | | | |
| 16 | | Toggle | | | | | | | | | | | | |
| 17 | | Toggle | | | | | | | | | | | | |
| 18 | | | 0 | 1 | 1 | 0 | | | | | | | | |
| 19 | | Toggle | | | | | | | | | | | | |
| 20 | | Toggle | | | | | | | | | | | | |
| 21 | | | 0 | 1 | 1 | 1 | | | | | | | | |

**Table 9-12**   Data for Parallel In/Parallel Out 74HC194N Shift Register

*Table continues next page*

| Step | Register Data Condition | | | | | | | | | | | | | |
|---|---|---|---|---|---|---|---|---|---|---|---|---|---|---|
| | Switch Settings | | | | | | Register 1 | | | | Register 2 | | | |
| | S1 Clear | S2 Clock | D | C | B | A | X1 | X2 | X3 | X4 | X5 | X6 | X7 | X8 |
| 22 | | Toggle | | | | | | | | | | | | |
| 23 | | Toggle | | | | | | | | | | | | |
| 24 | | | 1 | 0 | 0 | 0 | | | | | | | | |
| 25 | | Toggle | | | | | | | | | | | | |
| 26 | | Toggle | | | | | | | | | | | | |
| 27 | | | 1 | 0 | 0 | 1 | | | | | | | | |
| 28 | | Toggle | | | | | | | | | | | | |
| 29 | | Toggle | | | | | | | | | | | | |
| 30 | | | 1 | 0 | 1 | 0 | | | | | | | | |
| 31 | | Toggle | | | | | | | | | | | | |
| 32 | | Toggle | | | | | | | | | | | | |
| 33 | | | 1 | 0 | 1 | 1 | | | | | | | | |
| 34 | | Toggle | | | | | | | | | | | | |
| 35 | | Toggle | | | | | | | | | | | | |
| 36 | | | 1 | 1 | 0 | 0 | | | | | | | | |
| 37 | | Toggle | | | | | | | | | | | | |
| 38 | | Toggle | | | | | | | | | | | | |
| 39 | | | 1 | 1 | 0 | 1 | | | | | | | | |
| 40 | | Toggle | | | | | | | | | | | | |
| 41 | | Toggle | | | | | | | | | | | | |
| 42 | | | 1 | 1 | 1 | 0 | | | | | | | | |
| 43 | | Toggle | | | | | | | | | | | | |
| 44 | | Toggle | | | | | | | | | | | | |
| 45 | | | 1 | 1 | 1 | 1 | | | | | | | | |
| 46 | | Toggle | | | | | | | | | | | | |
| 47 | | Toggle | | | | | | | | | | | | |
| 48 | Down | | | | | | | | | | | | | |

**Table 9-12**   (continued)

### ● *Troubleshooting Problems:*

4. Open circuits file **d09-04c**. This circuit has three individual faults. The fault conditions are similar in each case. The visible problems are

_____

_____

and are caused by_____

_____

_____.

5. Open circuits file **d09-04d**. This circuit has only one fault. The visible

problem is _____

_____

_____

_____

and is caused by _____

_____

_____.

## ● *Circuit Construction:*

6. Open circuits file **d09-04e**. Use the components on the workspace to construct a parallel in/parallel out shift register according to Figure 9-11.

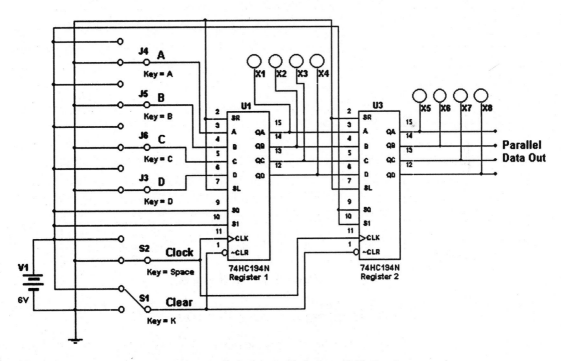

**Figure 9-11**  Parallel In/Parallel Out Shift Registers to be
Constructed with 74HC194N ICs

# 10. Schmitt Trigger, One-Shot and Clock Circuits

**References**

*MultiSIM* 2001

*MultiSIM* 2001 Study Guide

**Objectives**   After completing this chapter, you should be able to:

- Understand the input characteristics of Schmitt trigger circuits.
- Use the Schmitt Trigger circuit to develop a square wave clock signal.
- Check the frequency and operation of Schmitt Trigger circuits.
- Check the frequency and operation of 555 astable circuits.
- Check the frequency and operation of 555 monostable circuits.
- Use one-shot multivibrator circuits.

## Introduction

Advanced and specialized circuits such as Schmitt Trigger circuits, One-Shot (monostable) circuits, and timer circuits using devices such as the 555 timer IC provide for special applications that extend beyond the normal capabilities of other digital circuits. Technologies such as Schmitt trigger devices extend the realm of digital circuits and enhance their abilities, helping to speed up and shape inputs of older and slower integrated circuits.

## Activity 10.1: Schmitt Trigger Circuits for Wave Shaping and Clocks

1. One of the characteristics of integrated circuits using **Schmitt Trigger** technology is the ability to provide a reliable and accurate response to slow-rising, slowly changing waveforms at the input terminals of these circuits. This type of circuit ignores the input waveform until it reaches a certain threshold voltage point. At that point the circuit rapidly turns on (switches from a LOW condition to a HIGH condition). In the opposite direction, when the input voltage goes below the threshold voltage, the circuit switches to the opposite condition (LOW). The LOW threshold is less than the HIGH threshold. This differential between the two thresholds is entitled **hysteresis**.

2. The symbol used to indicate that an integrated circuit is of the Schmitt Trigger type is the hysteresis symbol ⎍ that you will see displayed on the logic symbol representing a 74LS14N digital circuit in Figure 10-1 and other Schmitt trigger circuits.

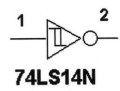

**74LS14N**

**Figure 10-1**  The Hysteresis Symbol Indicating a Schmitt Trigger Device

3. Open circuits file **d10-01a**. In this circuit, a 7414 Hex Inverter IC is being used to square up an input sine wave signal, turning it into a square wave. Activate the circuit and observe the input and output waveforms on the oscilloscope. Draw the input sine wave and the output square wave on Figure 10-2. What is the frequency of the square wave? The frequency of

the square wave is _____ Hz.

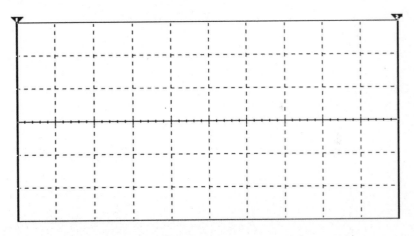

**Figure 10-2**  Input and Output Waveforms of 7414 Hex Inverter Circuit

4. Another use for Schmitt Trigger circuits is to develop timing and clock signals. Open circuits file **d10-01b**. In this circuit, the 74HC14N Hex Inverter IC is being used to produce a square wave. This type of circuit is called a relaxation oscillator and uses capacitor C1 and resistor R1 to determine the frequency of the output square wave. Activate the circuit and determine the frequency of the square wave. The frequency of the

square wave is _____ Hz. This type of circuit is used to develop timing signals within digital equipment. Draw the output waveform of the circuit on Figure 10-3 on page 134.

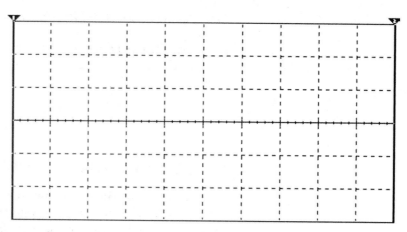

**Figure 10-3**   Output Waveform of 74HC14N Hex Inverter Circuit

5. Open circuits file **d10-01c**. In this circuit, the CMOS 40106BP Inverter IC is being used to develop a square wave. This circuit is similar to the circuit of Step 3 except that it uses a CMOS IC instead of a TTL IC. What is the frequency of the output square wave? The frequency of the output square wave is _____ Hz. This circuit also uses two output transistors to increase the power capabilities of the circuit. Draw the output waveform on Figure 10-4.

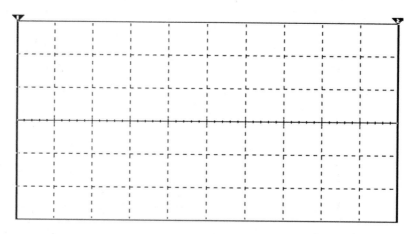

**Figure 10-4**   Output Waveform of 40416 Inverter Circuit

● *Troubleshooting Problems:*

6. Open circuits file **d10-01d**. This timing circuit is not working correctly; there is no output. What is wrong? The problem is _____

_____

_____

_____.

7. Open circuits file **d10-01e**. This wave shaping circuit is not working correctly; there is no output. What is wrong? The problem is _____

_____

_____.

### ● *Circuit Construction:*

8. Open circuits file **d10-01f**. Use the components on the workspace to construct an oscillator circuit. Draw your constructed circuit on Figure 10-5. What is the output frequency of your circuit? The output frequency of the

circuit is _____ Hz.

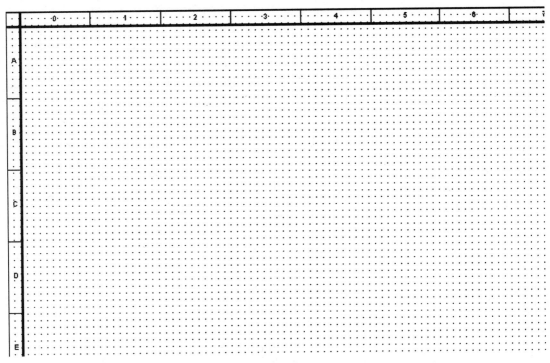

**Figure 10-5**   Constructed Oscillator Circuit Using Inverters

# Activity 10.2: The 555 Integrated Circuit as an Astable Multivibrator

1. The 555 IC is an integrated circuit that can be connected as an astable multivibrator (for a clock signal source) or used for other timing purposes in many types of analog and digital equipment. This circuit internally combines analog and digital circuits to produce timing signals in conjunction with external components that determine the output signal frequencies and types of waveforms.

2. The 555 IC has a trigger level equal to approximately one-third of the supply voltage and a threshold level equal to approximately two-thirds of the supply voltage. These levels can be altered by means of the control voltage terminal (CONT). When the trigger input (TRIG) falls below the trigger level, the flip-flop is set and the output goes HIGH. If TRIG is above the trigger level and the threshold input (THRES) is above the threshold level, the flip-flop is reset and the output is LOW. The reset input (RESET) can override all other inputs and can be used to initiate a new timing cycle. If RESET is LOW, the flip-flop is reset and the output is LOW. Whenever the output is LOW, a LOW impedance path is provided between the discharge terminal (DISCH) and GND. All unused inputs should be tied to an appropriate logic level to prevent false triggering. The pin-out diagram for a LM555N integrated circuit is displayed in Figure 10-6.

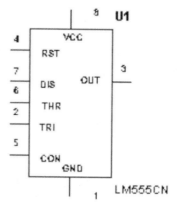

**Figure 10-6**    The LM555N Integrated Circuit

3. Open circuits file **d10-02a** and determine the frequency of the square

   wave. The frequency of the observed square wave is _____ Hz. Draw the observed output (Pin 3) waveform on Figure 10-7. What is the positive pulse width? The positive portion of the waveform is _____ milliseconds in width.

4. Open circuits file **d10-02b**. This circuit uses the output of a 555 timer circuit as a clock input for a 74LS73N flip-flop. What is the output frequency of the flip-flop? The output frequency of the flip-flop is

   _____ Hz. Draw the output waveform of the flip-flop on Figure 10-8.

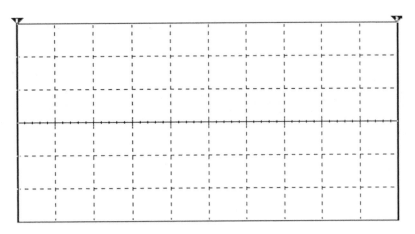

**Figure 10-7**   The Output Waveform of a 555 Timer Circuit

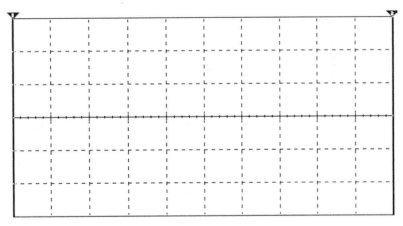

**Figure 10-8**   The Output Waveform of Timer/Flip-Flop Circuit

## ● *Troubleshooting Problem:*

5. Open circuits file **d10-02c**. In this circuit, the $Q$ output of the 74LS73 is missing. What is wrong? The problem is that _____

_____

_____.

## ● *Circuit Construction:*

6. Open circuits file **d10-02d**. Using the components on the workspace, construct a timer/flip-flop circuit similar to Figure 10-9. What is the frequency of the flip-flop with the potentiometer set for 95%, 50%, and 5%? The output frequency of the flip-flop at a setting of 95% is

_____ Hz, at 50% is _____ Hz, and at 5% is _____ Hz.

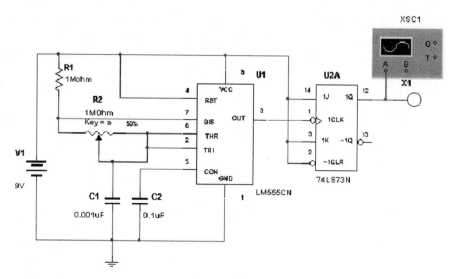

**Figure 10-9**  Timer/Flip-Flop Circuit to be Constructed

## Activity 10.3: The 555 Integrated Circuit as a Monostable Multivibrator

1. The 555 IC can also be used as a monostable multivibrator. This circuit produces a single output after being triggered into operation by an external trigger command on Pin 2 of the IC. The input (Pin 2) is normally held HIGH and activates circuit operation with a positive-to-negative transition of voltage. When the input is triggered, the output goes positive and the internal discharge-clamp transistor is released from the timing capacitor. This capacitor then recharges (positive) until two-thirds of the supply voltage is reached. At that instant, the internal threshold comparator flips the circuit and the output goes back to ground, a positive-to-negative transition. The circuit then remains in this state until the input is triggered into operation again. The circuit ignores any input changes until the timing cycle is complete. A typical monostable multivibrator circuit using the 555 IC is displayed in Figure 10-10.

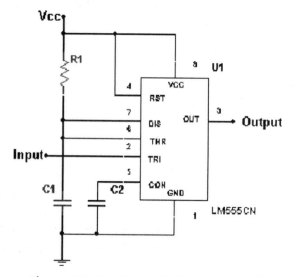

**Figure 10-10**  Typical 555 Monostable
Multivibrator Circuit

2. Open circuits file **d10-03a**. This circuit provides an output pulse of fixed duration. Activate the circuit and toggle input switch S1 (down and up). Did you notice that the output pulse is independent of the action of the input switch after the initial activation? Yes _____ No _____. What is the duration of the output pulse? The positive portion of the output pulse is _____ milliseconds in duration. What happens if the down-up motion of the switch is not completed, if the switch remains down? If the down-up motion of the switch is not completed, the output _____

_____

_____.

3. Open circuits file **d10-03b**. This circuit is an example of how a mono-stable multivibrator ignores the input signal until its timing cycle is complete. How many complete cycles of the input waveform occur during one complete cycle of the output waveform? _____ cycle(s) of the input waveform occurring during one complete cycle of the output waveform.

4. What is the duration of the positive portion of this output waveform? The output pulse is _____ milliseconds in duration.

## ● *Troubleshooting Problem:*

5. Open circuits file **d10-03c**. In this circuit, the output pulse is supposed to be about 11 milliseconds in duration. Instead the output pulse closely follows the toggle action of the input switch. There is very little delay from initial actuation to the end of the pulse. What is wrong? The problem is _____

_____

_____.

## ● *Circuit Construction:*

6. Open circuits file **d10-03d**. Using the components on the workspace, construct a circuit similar to Figure 10-11. What is the duration of the output (positive) pulse? The duration of the output pulse is _____ mSec. What is the frequency of the output waveform?

The frequency of the output waveform is _____ Hz. Draw the waveforms displayed on the oscilloscope on Figure 10-12.

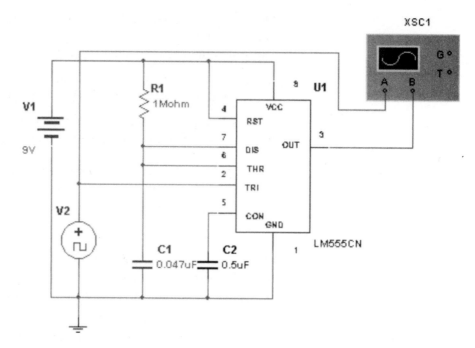

**Figure 10-11**   555 Timer Circuit to be Constructed

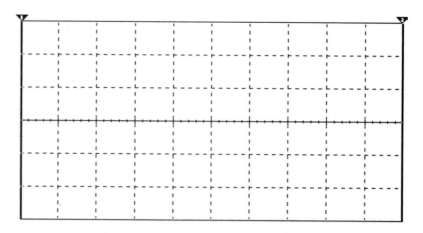

**Figure 10-12**   Waveforms of Constructed 555 Timer Circuit

## Activity 10.4:  One-Shot Circuits

1. **One-Shot** (OS) circuits can also be classified as monostable multivibrators. In a manner similar to the monostable 555 circuit, these circuits have one stable state or logic condition. They receive an actuation pulse, go HIGH and remain HIGH until the desired pulse period is complete as determined by external components.

2. One common usage of monostable circuits is to debounce switch inputs. Most switches produce quite a bit of contact bounce when they are actuated. Each individual bounce can cause problems at the input of a digital circuit because the circuit will tend to interpret each bounce as an individual input; thus, one input becomes multiple inputs.

3. Open circuits file **d10-04a**. This switch debounce circuit provides a single output for each switch activation by latching the monostable circuit ON until the pulse timing cycle is complete. Toggle the switch and observe the resultant waveform at the output of the monostable circuit. Was the output a pulse? Yes _____ No _____. This circuit is designed so that the output has only a small amount of delay. What is the duration of the pulse? The pulse is _____ milliseconds in duration.

4. Open circuits file **d10-04b**. This debounce circuit is similar to the previous circuit but has a longer delay after the input switch is toggled. What is the duration of the pulse? The output pulse is _____ milliseconds in duration.

### ● *Troubleshooting Problem:*

5. Open circuits file **d10-04c**. This debounce circuit isn't working correctly; the output remains HIGH. What is the problem? The problem is

_____

_____

_____.

### ● *Circuit Construction:*

6. Open circuits file **d10-04d**. Using the components on the workspace, construct a monostable circuit. Draw your circuit on Figure 10-13.

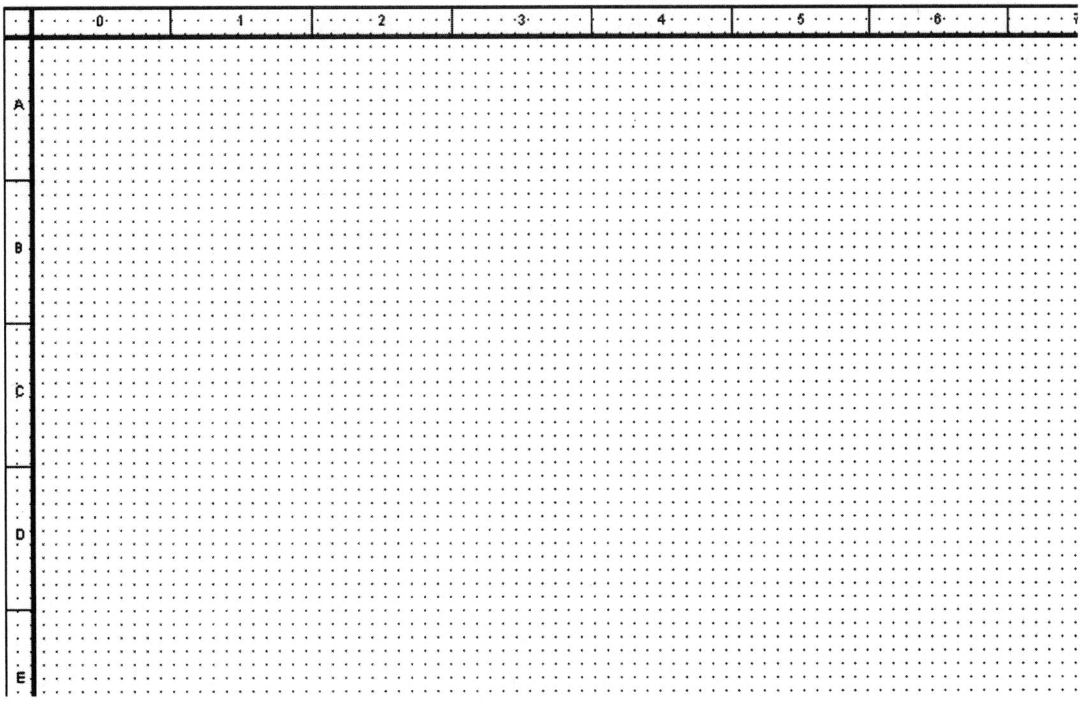

**Figure 10-13**   Constructed Monostable Circuit

## Activity 10.5: Crystal Oscillator Clock Circuits

1. When a circuit needs timing accuracy beyond the capabilities of an IC timer or a 555 timing circuit, a crystal oscillator is commonly used to provide needed stability and very precise timing. Many times a crystal device is used in conjunction with an integrated circuit such as the 4001 CMOS NOR gate to provide stable timing up to the frequency limits placed on the IC by internal propagation delay.

2. Open circuits file **d10-05a**. This is a typical clock circuit using a crystal for timing. What is the output frequency of the circuit? The output frequency of the circuit is _____ MHz. Draw the output waveform on Figure 10-14.

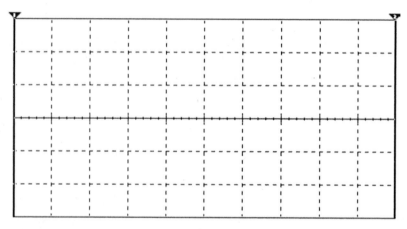

**Figure 10-14**   Waveform Produced by Crystal Clock Circuit

## ● *Troubleshooting Problem:*

3. Open circuits file **d10-05b**. This oscillator circuit is not working correctly, there does not appear to be any output. What is the problem? The

problem is _____

_____

_____.

### ● *Circuit Construction:*

4. Open circuits file **d10-05c**. Using the components on the workspace, construct a crystal oscillator circuit similar to the circuit of Figure 10-15.

5. Measure the frequency of the circuit. The frequency of the circuit is

   about _____ MHz. Is the frequency correct? Yes _____ No _____.
   If No, what is wrong (in this case you need to make an "educated"

   guess)? The problem is most likely _____

   _____

   _____.

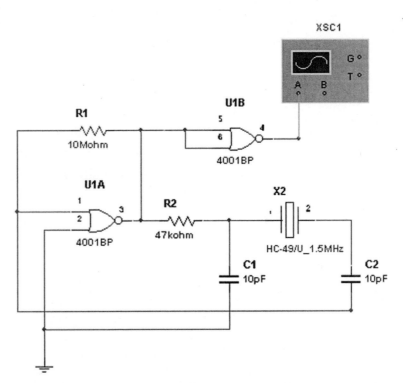

**Figure 10-15** Crystal Oscillator Circuit to be Constructed

# 11. Digital-to-Analog and Analog-to-Digital Converter Circuits

**References**

*MultiSIM* 2001

*MultiSIM* 2001 Study Guide

**Objectives**   After completing this chapter, you should be able to:

- Explain the operation of a digital-to-analog converter.
- Use resistor networks for digital-to-analog conversion.
- State the purpose of analog-to-digital conversion.
- State the purpose of digital-to-analog conversion.

## Introduction

Ultimately the digital world has to interface with the analog world. The concepts of time, space, temperature, pressure, voltage, light intensity, current, and many other measurable values are real world quantities that need to be digitally quantified before they can be used by digital circuits. Most physical variables, by nature are analog and can possibly have many values within a wide range of finite quantities. In turn, the logical actions/reactions of a digital system need to communicate its deterministic data with the analog environment.

A complete system of measurement, quantification, decision-making, and control might consist of the following:

- The measuring component (transducer)

- The **Analog-to-Digital Converter** (**ADC**)

- The digital computing device (decision-making)

- The **Digital-to-Analog Converter** (**DAC**)

- The controlling mechanism (actuator)

Analog-to-digital and digital-to-analog converters function as interface circuits operating between analog data and the digital system. A modern unit of equipment used in many industrial applications is the programmable logic controller (PLC). The PLC monitors the industrial environment through deterministic inputs (digital) or converted analog inputs. Then logical decisions are made based upon the conditions of the inputs and the logic of the previously installed operating program. Finally, the PLC provides outputs that communicate the logical decisions back to the analog world through various output devices.

## Activity 11.1: Digital-to-Analog Conversion: Voltage Output

1. Digital-to-Analog conversion generally consists of the process of taking a digital quantity (code) and converting it into a proportional voltage or current. A digital-to-analog (DAC) circuit can consist of discrete components or can be in the form of an integrated circuit with peripheral components to support its function. Several concerns that need to be addressed when using DACs are:

   - Output offset—the output level (voltage or current) of a DAC when all inputs are LOW. Ideally, the output offset is zero volts or zero amps when all of the inputs are LOW.

   - Resolution—the smallest change that can incrementally change the DAC output as the result of a minimum change in the digital input. Resolution is usually specified by DAC manufacturers in the number of bits (input) in relationship to the full-scale output capabilities of the device. To determine resolution, divide the maximum number of possible input switch combinations (255 for an eight-switch input) into the maximum output voltage.

2. Open circuits file **d11-01a**. In this circuit, eight switches provide digital inputs to a VDAC (V for voltage out). This type of DAC outputs a proportional voltage that ranges from 0 V to a maximum voltage. The maximum voltage output is adjustable by means of R9. If all of the switches are set LOW, then the output should be zero volts. If all the switches were set HIGH, then the output would be the maximum source voltage. In the MultiSIM program, this simulation, with a resultant voltage reading on the voltmeter, may take several minutes. Be patient. The switch number corresponds to the keyboard number; for example, S1 = 1 on the keyboard.

3. What is the resolution of this VDAC (in volts)? The resolution of this VDAC is (maximum voltage/255) _____ millivolts. In this circuit, the least significant digit (LSD) is input to the left (S0 to D0) of the VDAC, and the most significant digit (MSD) is input to the right (S7 to D7). Set the input switches to the position indicated by Table 11-1 and record the output voltage. Did you notice that Step 2 is equal to the resolution? Yes _____ No _____. Did you notice that Step 10 is the maximum output? Yes _____ No _____. It is always possible to calculate the resolution by dividing the output voltage

by the binary count (of the switches); for example, 0.3 V output divided by a binary count (closed switches) of 15 gives a resolution of 20 mV.

| | Input Switch Settings | | | | | | | | Output Voltage |
|---|---|---|---|---|---|---|---|---|---|
| Step | S7 | S6 | S5 | S4 | S3 | S2 | S1 | S0 | |
| 1 | 0 | 0 | 0 | 0 | 0 | 0 | 0 | 0 | |
| 2 | 0 | 0 | 0 | 0 | 0 | 0 | 0 | 1 | |
| 3 | 0 | 0 | 0 | 1 | 1 | 1 | 0 | 0 | |
| 4 | 0 | 0 | 1 | 0 | 1 | 1 | 1 | 1 | |
| 5 | 0 | 1 | 0 | 0 | 0 | 1 | 1 | 1 | |
| 6 | 1 | 0 | 0 | 0 | 1 | 1 | 0 | 0 | |
| 7 | 1 | 0 | 0 | 1 | 0 | 0 | 0 | 1 | |
| 8 | 1 | 0 | 1 | 0 | 0 | 0 | 0 | 0 | |
| 9 | 1 | 1 | 1 | 0 | 0 | 0 | 0 | 1 | |
| 10 | 1 | 1 | 1 | 1 | 1 | 1 | 1 | 1 | |

**Table 11-1**   VDAC Output Voltages

4. Open circuits file **d11-01b**. Your task with this circuit is to adjust the resolution (R9) until the maximum output (255 · resolution) is as close to 3 V as possible. Remember that the adjustment procedure consists of adjusting maximum output with R9. After your adjustment, what is the resolution of this circuit? The adjusted resolution of this circuit is

_____ millivolts. What is the maximum voltage output of this circuit?

The maximum voltage output of this circuit is _____ volts.

5. Open circuits file **d11-01c**. In this circuit, only four inputs are being used. The DMM in the circuit is connected to measure the VDAC circuit output. The 4040BP IC counter circuit is providing a continuously ascending binary count to the VDAC inputs. Open the clock input switch (K) connected to the function generator and to ~CP on the IC. Close the master reset (MR) switch (spacebar); this should place all of the output

indicators in the off condition. Did it? Yes _____ No _____.

6. Open the master reset switch (spacebar). Close the clock input switch (K) to allow the counter to count upwards until the four output indicators are turned on (X1–X4) and then open the switch. It may take several tries to accomplish this task. When all four indicators are on, the DMM will display the maximum output voltage of the VDAC. What is the maximum output voltage as displayed on the DMM? The maximum output voltage

is _____ millivolts. Use this figure to calculate circuit resolution. What is the voltage resolution of this circuit (maximum voltage/15)? The calculated voltage resolution of this circuit is

_____ millivolts. According to what we have learned thus far, the

amplifier output voltage should be _____V (resolution × number of steps).

7. Open and close the clock input switch (K) until only X1 is on. The display on the DMM should indicate circuit resolution. What is the voltage

reading? The output voltage reading is _____ millivolts. Did this gree with your calculation for circuit resolution in Step 6?

Yes _____ No _____.

8. This type of circuit uses the inherent capabilities of the VDAC to provide a specific voltage output for a specific digital input. If all eight inputs were being used, the output (on the oscilloscope) would resemble an ascending staircase with each step having a rise (in voltage) equal to the amount of the resolution of the circuit (255 steps). In this circuit, there are 15 steps with each step rising upwards by the amount of the resolution voltage calculated in Step 6 and measured in Step 7. This type of circuit is called a staircase generator.

9. Check the output of the circuit with the oscilloscope. There are

_____ steps in the staircase. What is the total height, in volts, of the staircase (this is the maximum voltage)? The height of the staircase is

proximately _____ volts. Does this agree with the maximum voltage

calculated in Step 6? Yes _____ No _____. Your answer should be "No." The reason is that the amplifier has gain which throws the voltage level off unless the gain is known. We will not expand this subject.

10. Expand the display on the oscilloscope by changing the Timebase scale to 1 ms/Div. What is the duration, in time (width), of each step? The

duration of each step is _____ millisecond(s).

## ● *Troubleshooting Problems:*

11. Open circuits file **d11-01d**. In this circuit, resolution was set for 17.55 millivolts with a maximum output of 4.475 volts. Obviously something is wrong, the voltage output is LOW. What is wrong? The problem is

_____

_____

_____.

12. Open circuits file **d11-01e**. In this circuit, the output seems to be incorrect and does not change when the inputs change. What is wrong? The

problem is _____

_____

_____.

## ● *Circuit Construction:*

13. Open circuits file **d11-01f**. Use the components on the workspace to construct a five-output DAC circuit. Set your maximum voltage to 2.5 V and determine the resultant resolution. The resolution would be

    _____ mV. Draw the circuit on Figure 11-1.

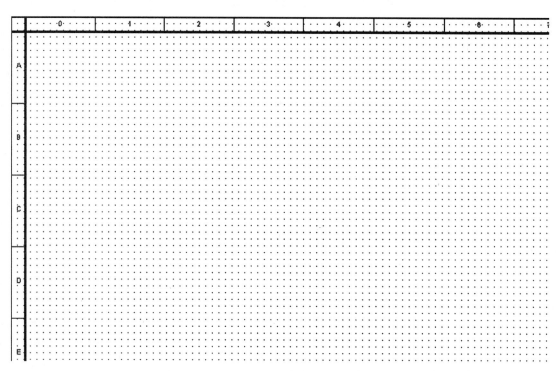

**Figure 11-1**   Constructed DAC Circuit

## Activity 11.2:  Digital-to-Analog Conversion: Current Output

1. Digital-to-Analog conversion with current output is similar to digital-to-analog conversion with voltage output, the only real difference being the type of output.

2. Open circuits file **d11-02a**. In this circuit, eight switches provide digital inputs to a IDAC (I for current out). This type of DAC provides an output that is a proportional current that ranges from 0 A to a maximum current. The maximum current output is adjustable by means of the reference voltage setting to the left of the device (R9). If all of the switches are set LOW, then the output should be zero amperes. If all the switches were set HIGH, then the output would rise to the maximum current. In

the MultiSIM program, this simulation with a resultant current reading on the ammeter may take some time, but not as long as a VDAC. Be patient.

3. What is the resolution of this IDAC (in microamperes)? The resolution

   of this IDAC is (maximum current/255) _____ milliamperes. In this circuit, the least significant digit (LSD) is input to the left (S0 to D0) of the IDAC and the most significant digit (MSD) is input to the right (S7 to D7). Set the input switches to the position indicated by Table 11-2 and record the output current as appropriate. Did you notice that Step 2 is

   equal to the resolution? Yes _____ No _____. Did you notice that

   Step 10 is the maximum current output? Yes _____ No _____.

| | Input Switch Settings | | | | | | | | Output Current |
|------|----|----|----|----|----|----|----|----|---|
| Step | S7 | S6 | S5 | S4 | S3 | S2 | S1 | S0 | |
| 1 | 0 | 0 | 0 | 0 | 0 | 0 | 0 | 0 | |
| 2 | 0 | 0 | 0 | 0 | 0 | 0 | 0 | 1 | |
| 3 | 0 | 0 | 0 | 0 | 1 | 1 | 0 | 0 | |
| 4 | 0 | 0 | 1 | 0 | 1 | 1 | 1 | 1 | |
| 5 | 0 | 1 | 0 | 0 | 0 | 1 | 1 | 1 | |
| 6 | 1 | 0 | 0 | 0 | 1 | 1 | 0 | 0 | |
| 7 | 1 | 0 | 0 | 1 | 0 | 0 | 0 | 1 | |
| 8 | 1 | 0 | 1 | 0 | 0 | 0 | 0 | 0 | |
| 9 | 1 | 1 | 1 | 0 | 0 | 0 | 0 | 1 | |
| 10 | 1 | 1 | 1 | 1 | 1 | 1 | 1 | 1 | |

**Table 11-2**   IDAC Output Currents

4. Open circuits file **d11-02b**. Your task with this circuit is to adjust the resolution (R9) until the maximum output current (255 · resolution) is as close to 100 milliamperes as possible. After your adjustment, what is the resolution of this circuit? The adjusted current resolution of this circuit is

   _____ microamperes. What is the maximum current output of this

   circuit? The maximum current output of this circuit is _____ mA.

5. Open circuits file **d11-02c**. In this circuit, only four inputs are being used. The DMM in the circuit is connected to measure the IDAC current output. The 4040BP IC counter circuit is providing a continuously ascending binary count to the IDAC inputs. Open the clock input switch (K) connected to the function generator and to ~CP on the IC. Close the

master reset (MR) switch (spacebar); this should place all of the output

indicators in the off condition. Did it? Yes _____ No _____.

6. Open the master reset switch (spacebar). Close the clock input switch (K) to allow the counter to count upwards until the four output indicators are turned on (X1–X4) and then open the switch. It may take several tries to accomplish this task. When all four indicators are on, the DMM will display the maximum output current of the IDAC. What is the maximum output current as displayed on the DMM? The maximum output current

is _____ milliamperes. Use this figure to calculate circuit resolution. What is the resolution of this circuit (maximum current/15)? The

calculated resolution of this IDAC circuit is _____ microamperes.

7. Open and close the clock input switch (K) until only X1 is on. The display on the DMM should indicate current resolution. What is the cur-

rent reading? The output current reading is _____ microamperes. Did this agree with your calculation for circuit resolution in Step 6?

Yes _____ No _____.

8. This type of circuit uses the inherent capabilities of the IDAC to provide a specific current output for a specific digital input. If all eight inputs were being used, the output (on the oscilloscope) would resemble an ascending staircase with each step having a rise (in current) equal to the amount of the resolution of the circuit (255 steps). In this circuit, there are 15 steps with each step rising upwards by the amount of the resolution current calculated in Step 6 and measured in Step 7. This type of circuit is called a staircase generator, generating predictable current steps.

9. Check the output of the circuit with the oscilloscope. There are

_____ steps in the staircase. What is the total height, displayed in volts, of the staircase (representing the maximum current)? The height

of the staircase is approximately _____ volts representing milliamperes. Does this agree with the maximum current calculated in

Step 6? Yes _____ No _____.

10. Expand the display on the oscilloscope by changing the Timebase scale to 1 ms/Div. What is the duration, in time (width), of each step? The

duration of each step is _____ millisecond(s).

## ● *Troubleshooting Problems:*

11. Open circuits file **d11-02d**. In this circuit, current resolution was set for 32.5 μA with a maximum current output of 8.287 mA. Obviously something is wrong, there is no output. What is wrong? The problem is

    _____

    _____

    _____

    _____.

12. Open circuits file **d11-02e**. In this circuit, nothing is coming out of the IDAC, and it looks like the counter is not working. What is wrong? The

    problem is _____

    _____

    _____

    _____.

## ● *Circuit Construction:*

13. Open circuits file **d11-02f**. Use the components to construct an IDAC similar to Figure 11-2. Adjust the resolution to 32.5 μA. What is the maximum output? The maximum output is _____ mA.

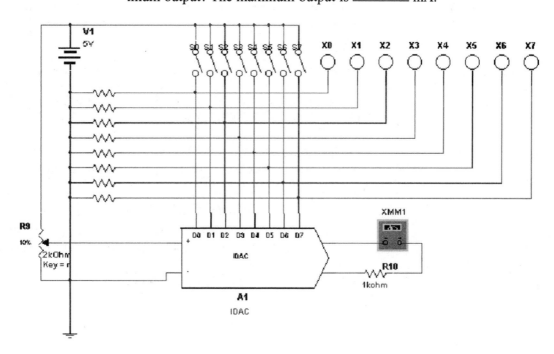

**Figure 11-2**   IDAC Circuit to be Constructed

## Activity 11.3: Analog-to-Digital Conversion

1. Analog-to-Digital conversion is similar to digital-to-analog conversion except that the opposite task is accomplished and the process is more complicated and time consuming. Instead of changing digital quantities into analog quantities, analog quantities are changed into digital quantities. There are many types of circuits, each with its own advantages that accomplish the task of changing an infinite array of analog data into discrete digital codes. We will look at one of these circuits.

2. Open circuits file **d11-03a**. In this circuit, the **digital ramp** technique is used for analog-to-digital conversion. As the input voltage is increased (using I), more and more of the output lamps should turn on. Activate the circuit, increase the input voltage according to the steps enumerated in Table 11-3, and enter the resultant data in the table. Mark an "X" under the lamp that is turned on for each voltage step. The voltage displayed on M1 is the input voltage and the voltage displayed on M2 is the reference voltage.

| Input Voltage | Output Indicators | | | | | | | |
|---|---|---|---|---|---|---|---|---|
| | X0 | X1 | X2 | X3 | X4 | X5 | X6 | X7 |
| 0 V | | | | | | | | |
| 0.5 V | | | | | | | | |
| 1.0 V | | | | | | | | |
| 1.5 V | | | | | | | | |
| 2.0 V | | | | | | | | |
| 2.5 V | | | | | | | | |
| 3.0 V | | | | | | | | |
| 3.5 V | | | | | | | | |
| 4.0 V | | | | | | | | |
| 4.5 V | | | | | | | | |
| 5.0 V | | | | | | | | |

**Table 11-3**   ADC Data Conversion

3. In Table 11-3, the binary value of X7 (regarding output data) is 128/255. At the 2.5 V input step, what is the binary value of the "ON" lamps? The

   binary value of the turned on lamps, X1, _____

   _____ (list the turned on lamps), is _____ /255. Return the input voltage to 0 V.

4. Return the input voltage to 0 V ( i ). Adjust the threshold voltage (a) to 3.0 V. Now activate the circuit and enter the new data into Table 11-4.

| Input Voltage | Output Indicators | | | | | | | |
|---|---|---|---|---|---|---|---|---|
| | X0 | X1 | X2 | X3 | X4 | X5 | X6 | X7 |
| 0 V | | | | | | | | |
| 0.5 V | | | | | | | | |
| 1.0 V | | | | | | | | |
| 1.5 V | | | | | | | | |
| 2.0 V | | | | | | | | |
| 2.5 V | | | | | | | | |

**Table 11-4**   ADC Data After Changing the Threshold Voltage

## ● *Troubleshooting Problem:*

5. Open circuits file **d11-03b**. This circuit is not simulating correctly. What

   is the problem? The problem is _____

   _____

   _____

   _____.

## ● *Circuit Construction:*

6. Open circuits file **d11-03c**. Using the components on the workspace, construct an ADC circuit similar to Figure 11-3. After the circuit is constructed, adjust M2 for 2.5 V. What is the resolution of the circuit? The

resolution is about _____ mV.

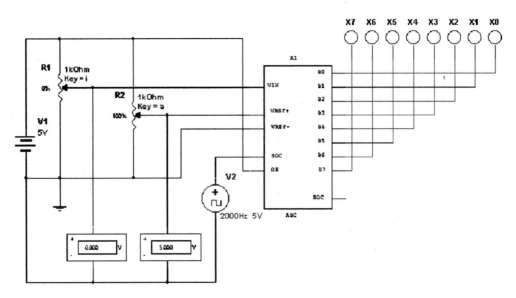

**Figure 11-3**   ADC Circuit to be Constructed

# 12. Decoders, Encoders, Code Converters, and Displays

> **References**
> *MultiSIM* 2001
> *MultiSIM* 2001 Study Guide

**Objectives**   After completing this chapter, you should be able to:

- Analyze and use decoder and encoder circuits.
- Determine logical outputs of decoder circuits.
- Determine logical outputs of encoder circuits.
- Connect display devices and assemblies.

## Introduction

Digital systems continually use digital circuits that operate, in various ways, on encoded binary messages. These types of circuits fall in the combinational category of digital circuits and include:

- *Decoders and Encoders:* circuits that detect and determine the binary state of input messages and change that message to a different output code.

- *Code Converters:* circuits that convert a coded binary message into another binary code.

Many of these combinational circuits are further categorized as medium-scale integration (MSI) logic devices. There are many relationships between these circuits and the circuits that have been studied in the preceding chapters.

One of the most common uses of coders, decoders, and code converters is to interface the data output of a digital system with some type of display such as a 7-segment display, an LCD display unit, or one of many other types of indicators.

## Activity 12.1: Binary-to-Decimal Decoder Circuits

1. **Binary-to-Decimal Decoder** circuits translate binary input data, representing specific digital information, into signals that energize corresponding outputs. The simplest decoder circuit consists of a single AND or NAND gate that provides an output representing a specific numerical count when all of its inputs are HIGH. A simple AND gate that decodes a three-input binary code and detects the decimal number 6 is displayed in Figure 12-1. Notice that the Boolean equation for the circuit is $\overline{A}BC = 6$.

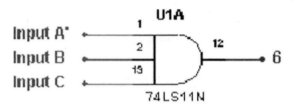

**Figure 12-1** AND Gate "6" Decoder Circuit

2. Open circuits file **d12-01a**. In this circuit, AND gates and inverters are used to decode a three-input binary code into eight possible outputs representing the decimal numbers 0 through 7. This circuit is referred to as a 1-of-8 decoder. Activate the circuit according to Table 12-1 and enter the resulting data into the table.

| Inputs | | | Outputs | | | | | | | |
|---|---|---|---|---|---|---|---|---|---|---|
| C | B | A | X7 | X6 | X5 | X4 | X3 | X2 | X1 | X0 |
| 0 | 0 | 0 | | | | | | | | |
| 0 | 0 | 1 | | | | | | | | |
| 0 | 1 | 0 | | | | | | | | |
| 0 | 1 | 1 | | | | | | | | |
| 1 | 0 | 0 | | | | | | | | |
| 1 | 0 | 1 | | | | | | | | |
| 1 | 1 | 0 | | | | | | | | |
| 1 | 1 | 1 | | | | | | | | |

**Table 12-1** 1-of-8 Decoder Data

3. Open circuits file **d12-01b**. This circuit is similar to the circuit of Step 2 except that all of the outputs are normally HIGH if the IC is not enabled. When enabled, the output goes LOW representing a specific number between 0 and 7. These enable inputs are G1 and ~G2A or ~G2B. The circuit enable Boolean equation is G1 · (~G2A + ~G2B) with the OR part of the equation indicating that either ~G2A or ~G2B (or both) has to be

connected LOW with the unused input floating. If ~G2A or ~G2B is connected HIGH the circuit will not be enabled. This circuit is designated as a 3-Line to 8-Line Decoder/Demultiplexer. Activate the circuit according to Table 12-2 and enter the resulting data into the table.

| Inputs | | | | | Outputs | | | | | | | |
|---|---|---|---|---|---|---|---|---|---|---|---|---|
| G1 | ~G2A + ~G2B | C | B | A | X7 | X6 | X5 | X4 | X3 | X2 | X1 | X0 |
| X | H | X | X | X | 1 | 1 | 1 | 1 | 1 | 1 | 1 | 1 |
| L | X | X | X | X | 1 | 1 | 1 | 1 | 1 | 1 | 1 | 1 |
| H | L | 0 | 0 | 0 | | | | | | | | |
| H | L | 0 | 0 | 1 | | | | | | | | |
| H | L | 0 | 1 | 0 | | | | | | | | |
| H | L | 0 | 1 | 1 | | | | | | | | |
| H | L | 1 | 0 | 0 | | | | | | | | |
| H | L | 1 | 0 | 1 | | | | | | | | |
| H | L | 1 | 1 | 0 | | | | | | | | |
| H | L | 1 | 1 | 1 | | | | | | | | |

**Table 12-2**   3-Line to 8-Line Decoder/Demultiplexer Data

## ● *Troubleshooting Problems:*

4. Open circuits file **d12-01c**. This circuit has a circuit problem; output X3 will not change states. What is wrong? The problem is _____

_____

_____

_____.

5. Open circuits file **d12-01d**. This circuit has a connection problem; the outputs stay HIGH, never going LOW. What is wrong? The problem is

_____

_____

_____

_____.

● *Circuit Construction:*

6. Open circuits file **d12-01e**. Using the components and circuit on the workspace, construct a circuit that will detect the numbers 1, 3, 5, or 7. Activate the circuit and enter the resultant data in Table 12-3. The output indicator X8 should illuminate when one of the required numbers is detected.

| Inputs | | | Output |
|---|---|---|---|
| **C** | **B** | **A** | **X8** |
| 0 | 0 | 0 | |
| 0 | 0 | 1 | |
| 0 | 1 | 0 | |
| 0 | 1 | 1 | |
| 1 | 0 | 0 | |
| 1 | 0 | 1 | |
| 1 | 1 | 0 | |
| 1 | 1 | 1 | |

**Table 12-3**   Data Table for Constructed Decoder Circuit

## Activity 12.2: BCD-to-Decimal Decoder Circuits

1. **BCD-to-Decimal Decoder** circuits translate binary input data representing a Binary Coded Decimal (BCD) code into signals that energize corresponding decimal outputs. A typical decoder IC is displayed in Figure 12-2. The BCD inputs are on the right side and the 7-segment outputs are on the left side. The three pins at the bottom left are enable inputs.

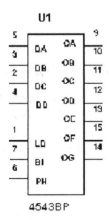

**Figure 12-2**   4543 BP Decoder IC

2. Open circuits file **d12-01a**. This circuit uses the 74LS42N BCD-to-decimal decoder in a manner similar to the 4543BP circuit of Step 2. Activate the circuit and record the resultant data in Table 12-4. Notice that the input data does not go beyond a binary count of 9.

| Inputs | | | | Outputs | | | | | | | | | |
|---|---|---|---|---|---|---|---|---|---|---|---|---|---|
| D | C | B | A | X9 | X8 | X7 | X6 | X5 | X4 | X3 | X2 | X1 | X0 |
| 0 | 0 | 0 | 0 | | | | | | | | | | |
| 0 | 0 | 0 | 1 | | | | | | | | | | |
| 0 | 0 | 1 | 0 | | | | | | | | | | |
| 0 | 0 | 1 | 1 | | | | | | | | | | |
| 0 | 1 | 0 | 0 | | | | | | | | | | |
| 0 | 1 | 0 | 1 | | | | | | | | | | |
| 0 | 1 | 1 | 0 | | | | | | | | | | |
| 0 | 1 | 1 | 1 | | | | | | | | | | |
| 1 | 0 | 0 | 0 | | | | | | | | | | |
| 1 | 0 | 0 | 1 | | | | | | | | | | |

**Table 12-4**   74LS42 BCD-to-Decimal Decoder Circuit Data

3. What happens to the output of the circuit if all inputs are placed HIGH?

The output indicators _____.

Any time the input count goes higher than 9, the outputs _____

_____

_____

## ● *Troubleshooting Problems:*

4. Open circuits file **d12-02b**. This circuit has a connection problem, what is it? Activate the circuit and enter the resultant data in Table 12-5 on

page 160. The problem is _____

_____

_____

_____

| Inputs | | | | Outputs | | | | | | | | | |
|---|---|---|---|---|---|---|---|---|---|---|---|---|---|
| D | C | B | A | X9 | X8 | X7 | X6 | X5 | X4 | X3 | X2 | X1 | X0 |
| 0 | 0 | 0 | 0 | | | | | | | | | | |
| 0 | 0 | 0 | 1 | | | | | | | | | | |
| 0 | 0 | 1 | 0 | | | | | | | | | | |
| 0 | 0 | 1 | 1 | | | | | | | | | | |
| 0 | 1 | 0 | 0 | | | | | | | | | | |
| 0 | 1 | 0 | 1 | | | | | | | | | | |
| 0 | 1 | 1 | 0 | | | | | | | | | | |
| 0 | 1 | 1 | 1 | | | | | | | | | | |
| 1 | 0 | 0 | 0 | | | | | | | | | | |
| 1 | 0 | 0 | 1 | | | | | | | | | | |

**Table 12-5** Output Data for Defective 74LS42 Circuit

5. Open circuits file **d12-02c**. This circuit has a circuit problem, what is it?

The problem is _____

_____

_____

_____ .

## ● *Circuit Construction:*

6. Open circuits file **d12-02d**. Using the components and circuit on the workspace, construct a circuit that will illuminate a lamp (X11) when the input binary count is greater than 9. The method to solve the problem is to AND all of the outputs. If the input exceeds 9, all of the outputs will go HIGH.

## Activity 12.3: BCD-to-Seven-Segment Decoder/Driver Circuits and Seven-Segment Displays

1. **BCD-to-Seven-Segment Decoder/Driver** circuits translate binary input data representing a Binary Coded Decimal (BCD) code into signals that energize corresponding decimal outputs. The 7446N BCD-to-7-segment IC (open collector output) is displayed in Figure 12-3.

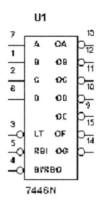

**Figure 12-3**  7446 BCD-to-Seven-Segment IC

2. Open circuits file **d12-03a**. This circuit uses the 7447 BCD-to-7-segment decoder/driver IC to translate the binary coded decimal signal into a 7-segment display. The OA through OG outputs activate the proper segments (a through g—see Figure 12-4) of the display unit to produce a decimal number. This is a common task for digital equipment, sending out signals to various types of display units. Activate the circuit and enter the resultant data into Table 12-6.

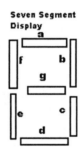

**Figure 12-4**    Segment Lettering of a typical
Seven-Segment Display

| Inputs | | | | Outputs |
|---|---|---|---|---|
| **D** | **C** | **B** | **A** | **Display** |
| 0 | 0 | 0 | 0 | |
| 0 | 0 | 0 | 1 | |
| 0 | 0 | 1 | 0 | |
| 0 | 0 | 1 | 1 | |
| 0 | 1 | 0 | 0 | |
| 0 | 1 | 0 | 1 | |
| 0 | 1 | 1 | 0 | |
| 0 | 1 | 1 | 1 | |
| 1 | 0 | 0 | 0 | |
| 1 | 0 | 0 | 1 | |

**Table 12-6**    BCD-to-Seven-Segment Display Data

## ● *Troubleshooting Problems:*

3. Open circuits file **d12-03b**. This circuit has a component problem. It displays incorrect decimal numbers on the 7-segment display. Activate the circuit and enter the resultant data into Table 12-7.

| Step | Inputs | | | | Outputs |
|:---:|:---:|:---:|:---:|:---:|:---:|
| | **D** | **C** | **B** | **A** | **Display** |
| 1 | 0 | 0 | 0 | 0 | |
| 2 | 0 | 0 | 0 | 1 | |
| 3 | 0 | 0 | 1 | 0 | |
| 4 | 0 | 0 | 1 | 1 | |
| 5 | 0 | 1 | 0 | 0 | |
| 6 | 0 | 1 | 0 | 1 | |
| 7 | 0 | 1 | 1 | 0 | |
| 8 | 0 | 1 | 1 | 1 | |
| 9 | 1 | 0 | 0 | 0 | |
| 10 | 1 | 0 | 0 | 1 | |

**Table 12-7**   Troubleshooting Data for Circuit **d12-03b**

4. Which steps have incorrect data? The decimal display data for Steps

_____ are incorrect. What

is wrong with the circuit? The problem with the circuit is that _____

_____

_____

_____ .


## ● *Circuit Construction:*

5. Open circuits file **d12-03c**. Using the components and the 7447N IC on the workspace, connect the 7-segment display to the outputs of the IC. Then construct a logic circuit that will detect a count of six (6) using the four-input AND gates. Activate the circuit and enter the resultant data in Table 12-8.

| Step | Inputs | | | | Outputs | |
|:---:|:---:|:---:|:---:|:---:|:---:|:---:|
| | **D** | **C** | **B** | **A** | **7-Segment Display** | **3-Detector Circuit** |
| 1 | 0 | 0 | 0 | 0 | | |
| 2 | 0 | 0 | 0 | 1 | | |
| 3 | 0 | 0 | 1 | 0 | | |
| 4 | 0 | 0 | 1 | 1 | | |
| 5 | 0 | 1 | 0 | 0 | | |
| 6 | 0 | 1 | 0 | 1 | | |
| 7 | 0 | 1 | 1 | 0 | | |
| 8 | 0 | 1 | 1 | 1 | | |
| 9 | 1 | 0 | 0 | 0 | | |
| 10 | 1 | 0 | 0 | 1 | | |

**Table 12-8**   Data for Constructed Six-Detector Circuit and 7-Segment Display Circuit

## Activity 12.4: Displaying Hexadecimal Numbers With Decoded Seven-Segment Displays

1. **Hexadecimal-to-Seven-Segment Decoder/Driver** circuits translate binary input data representing a hexadecimal number into signals that energize corresponding 7-segment display outputs. The circuit that is used for hexadecimal displays is similar to the circuit that is used for BCD-to-7-segment displays. The major difference is that the decoder is part of the display unit and is able to display hexadecimal numbers that extend beyond the decimal number 9 to the hexadecimal number F (decimal 15).

2. Open circuits file **d12-04a**. This circuit uses a decoded 7-segment decoder/driver display to translate the binary-coded hexadecimal signal into a 7-segment display. Activate the circuit and enter the resultant data into Table 12-9 on page 164.

| Inputs | | | | Outputs |
|---|---|---|---|---|
| D | C | B | A | Display |
| 0 | 0 | 0 | 0 | |
| 0 | 0 | 0 | 1 | |
| 0 | 0 | 1 | 0 | |
| 0 | 0 | 1 | 1 | |
| 0 | 1 | 0 | 0 | |
| 0 | 1 | 0 | 1 | |
| 0 | 1 | 1 | 0 | |
| 0 | 1 | 1 | 1 | |
| 1 | 0 | 0 | 0 | |
| 1 | 0 | 0 | 1 | |
| 1 | 0 | 1 | 0 | |
| 1 | 0 | 1 | 1 | |
| 1 | 1 | 0 | 0 | |
| 1 | 1 | 0 | 1 | |
| 1 | 1 | 1 | 0 | |
| 1 | 1 | 1 | 1 | |

**Table 12-9** Hexadecimal to Decoded Seven-Segment Display Data

● *Troubleshooting Problems:*

3. Open circuits file **d12-04b**. This circuit has connection problems that results in incorrect decimal data on the 7-segment display unit. Activate the circuit and enter the resultant data into Table 12-10.

4. Which steps have incorrect data? The decimal display data for Steps

_____ are incorrect. What

is wrong with the circuit? The problem with the circuit is that _____

_____

_____

_____ .

| Step | Inputs | | | | Outputs |
|:---:|:---:|:---:|:---:|:---:|:---:|
| | **D** | **C** | **B** | **A** | **Display** |
| 1 | 0 | 0 | 0 | 0 | |
| 2 | 0 | 0 | 0 | 1 | |
| 3 | 0 | 0 | 1 | 0 | |
| 4 | 0 | 0 | 1 | 1 | |
| 5 | 0 | 1 | 0 | 0 | |
| 6 | 0 | 1 | 0 | 1 | |
| 7 | 0 | 1 | 1 | 0 | |
| 8 | 0 | 1 | 1 | 1 | |
| 9 | 1 | 0 | 0 | 0 | |
| 10 | 1 | 0 | 0 | 1 | |
| 11 | 1 | 0 | 1 | 0 | |
| 12 | 1 | 0 | 1 | 1 | |
| 13 | 1 | 1 | 0 | 0 | |
| 14 | 1 | 1 | 0 | 1 | |
| 15 | 1 | 1 | 1 | 0 | |
| 16 | 1 | 1 | 1 | 1 | |

**Table 12-10**   Troubleshooting Data for Circuit **d12-04b**

## ● *Circuit Construction:*

5. Open circuits file **d12-04c**. Using the components on the workspace, connect the 7-segment display to the data switches. Then construct a logic circuit that will detect a hexadecimal count of b (11 in decimal) using the inverters and the AND gate. Activate the circuit and enter the resultant data in Table 12-11.

| Step | Inputs | | | | Outputs |
|:---:|:---:|:---:|:---:|:---:|:---:|
| | **D** | **C** | **B** | **A** | **Display** |
| 1 | 0 | 0 | 0 | 0 | |
| 2 | 0 | 0 | 0 | 1 | |
| 3 | 0 | 0 | 1 | 0 | |
| 4 | 0 | 0 | 1 | 1 | |

**Table 12-11**   Data for Constructed Three-Detector Circuit

*Table continues next page*

| Step | Inputs | | | | Outputs |
|---|---|---|---|---|---|
| | D | C | B | A | Display |
| 5 | 0 | 1 | 0 | 0 | |
| 6 | 0 | 1 | 0 | 1 | |
| 7 | 0 | 1 | 1 | 0 | |
| 8 | 0 | 1 | 1 | 1 | |
| 9 | 1 | 0 | 0 | 0 | |
| 10 | 1 | 0 | 0 | 1 | |
| 11 | 1 | 0 | 1 | 0 | |
| 12 | 1 | 0 | 1 | 1 | |
| 13 | 1 | 1 | 0 | 0 | |
| 14 | 1 | 1 | 0 | 1 | |
| 15 | 1 | 1 | 1 | 0 | |
| 16 | 1 | 1 | 1 | 1 | |

**Table 12-11**    (*continued*)

## Activity 12.5: Encoding—Decimal to BCD (Ten-to-Four-Line)

1. An **Encoder** circuit has a number of input lines, each of which has an individual decimal value. When one of the input lines becomes active, a binary output code related to the individual input is produced by the encoder circuit. The digital encoder is complementary to the digital decoder and works in the opposite direction, producing a particular digital code. One such type encoder is the **Decimal-to-BCD** circuit.

2. Open circuits file **d12-05a**. This circuit uses a 74LS147D 10-Line-to-4-Line Encoder IC to input decimal signals into BCD code. Activate the circuit and enter the resultant data into Table 12-12.

3. Open circuits file **d12-05b**. This circuit uses a 74LS147D 10-Line-to-4-Line Encoder IC to convert input decimal signals into BCD code and then send the output to a decoded 7-segment display circuit. Activate the circuit and enter the resultant data into Table 12-13.

| Decimal Input | BCD Data Output | | | |
|---|---|---|---|---|
| | D | C | B | A |
| 0 | | | | |
| 1 | | | | |
| 2 | | | | |
| 3 | | | | |
| 4 | | | | |
| 5 | | | | |
| 6 | | | | |
| 7 | | | | |
| 8 | | | | |
| 9 | | | | |

**Table 12-12**   Decimal-to-BCD Encoder Circuit Data

| Decimal Input | 7-Segment Data Output |
|---|---|
| 0 | |
| 1 | |
| 2 | |
| 3 | |
| 4 | |
| 5 | |
| 6 | |
| 7 | |
| 8 | |
| 9 | |

**Table 12-13**   Decimal-to-BCD Encoder Circuit Data

● *Troubleshooting Problem:*

4. Open circuits file **d12-04c**. This circuit has a circuit problem that causes outputs A and D to remain on. The input switches have no effect on the

   output data. What is the problem? The problem is _____

   _____

   _____

   _____ .

● *Circuit Construction:*

5. Open circuits file **d12-04d**. Using the components on the workspace, connect the 7-segment display to the output of the IC so that it can display the decimal value of the BCD code. Activate the circuit and record the resultant data in Table 12-14.

| Decimal Input | BCD Data Output | | | | 7-Segment Data |
|---|---|---|---|---|---|
| | D | C | B | A | |
| 0 | | | | | |
| 1 | | | | | |
| 2 | | | | | |
| 3 | | | | | |
| 4 | | | | | |
| 5 | | | | | |
| 6 | | | | | |
| 7 | | | | | |
| 8 | | | | | |
| 9 | | | | | |

**Table 12-14**  Decimal-to-BCD Encoder Circuit Data and Seven-Segment Display

# 13. More Combinational Circuits: Multiplexers and Demultiplexers

> **References**
> *MultiSIM* 2001
> *MultiSIM* 2001 Study Guide

**Objectives** After completing this chapter, you should be able to:

- Use MultiSIM to multiplex and demultiplex digital data.
- Use data selectors to multiplex and demultiplex digital information.
- Understand the operation of multiplexers and demultiplexers.
- Determine the content of a multiplexed message.
- Determine the mode and method necessary to transmit digital data from multiple sources over a single transmission line.

## Introduction

As the requirements and needs of digital logic have historically increased, basic logic gates and logic devices have been developed and married together into combinational circuits designed to meet those specific requirements and needs. These complex combinational circuits typically use similar methods to deal with certain aspects of digital data. One group of combinational circuits that operate together in a similar manner is designated as **Multiplexers** and **Demultiplexers**.

- Multiplexer (**MUX**) circuits are used to logically combine incoming binary data from multiple sources into a single outgoing data stream, yet retain the individuality of the source data in the outgoing message. They are also used to select transmission paths and modes for the outgoing data. The data is taken from a group of sources and transmitted on one outgoing data path. The result of the multiplex process is a multiplicity of incoming data sources and a single outgoing data transmission path.

- Demultiplexer (**DEMUX**) circuits are used to receive multiplexed data from one or more transmission paths and to restore that data into its original mode of multiple sources. A typical result is one stream of incoming data on a transmission line with multiple data paths out; the multiplexed data has been de-multiplexed.

# Activity 13.1: Multiplexing Digital Data

1. Multiplexer circuits range in complexity from basic combinational circuits to very complex ICs. They are used primarily for data routing—selecting a transmission path for outgoing data according to the coding criteria established by binary inputs.

2. Open circuits file **d13-01a**. This basic multiplexer circuit uses switches to determine which of the two signals will be transmitted on the output data line. Activate the circuit and determine the frequency of each of the signal sources. The frequency of signal source V 2 is _____ kHz and of signal source V 3 is _____ kHz. Draw the scope display on Figure 13-1 for signal source V 2 and then draw the scope display on Figure 13-2 for signal source V 3.

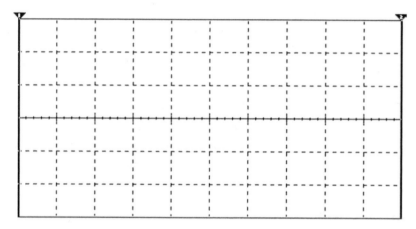

**Figure 13-1**   Scope Display of Multiplexer Circuit Signals for Signal Source V 2

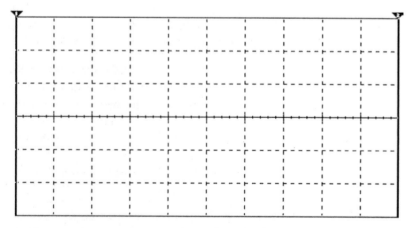

**Figure 13-2**   Scope Display of Multiplexer Circuit Signals for Signal Source V 3

3. Open circuits file **d13-01b**. In this circuit, we are going beyond the switch-controlled multiplexer circuit to a signal-controlled circuit. This is a true multiplexer circuit, the signal on the output line alternates between the input signal sources according to the controlling signal. Thus, there are three frequencies, the frequencies of the signal sources and also the frequency of the controlling signal. Activate the circuit and determine these three frequencies. The frequency of control signal V 1 is

_____ kHz, and of signal source V 2 is _____ kHz, and signal

source V 3 is _____ kHz. The top signal on the oscilloscope display is the controlling signal and the lower trace is the multiplexed signal. Draw the scope display and identify the multiplexed portions of the output signal that represent signal source V 2 and signal source V 3 on Figure 13-3. Signal source V 3 has a higher frequency than V 2.

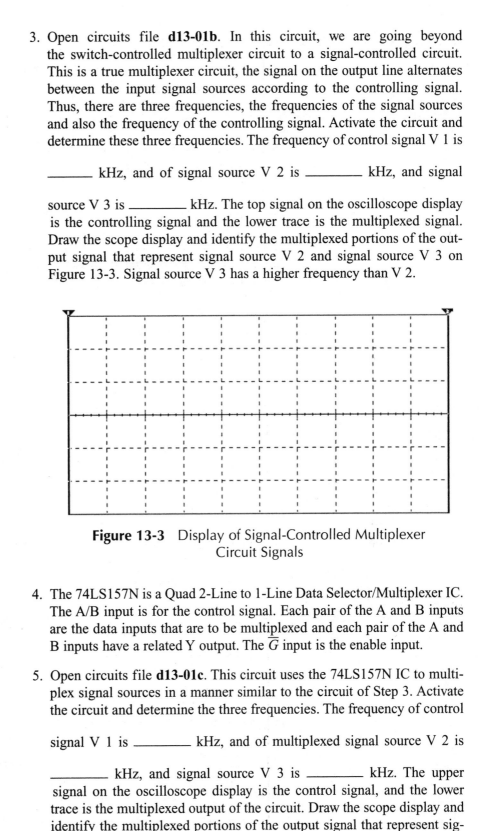

**Figure 13-3** Display of Signal-Controlled Multiplexer Circuit Signals

4. The 74LS157N is a Quad 2-Line to 1-Line Data Selector/Multiplexer IC. The A/B input is for the control signal. Each pair of the A and B inputs are the data inputs that are to be multiplexed and each pair of the A and B inputs have a related Y output. The $\overline{G}$ input is the enable input.

5. Open circuits file **d13-01c**. This circuit uses the 74LS157N IC to multiplex signal sources in a manner similar to the circuit of Step 3. Activate the circuit and determine the three frequencies. The frequency of control

signal V 1 is _____ kHz, and of multiplexed signal source V 2 is

_____ kHz, and signal source V 3 is _____ kHz. The upper signal on the oscilloscope display is the control signal, and the lower trace is the multiplexed output of the circuit. Draw the scope display and identify the multiplexed portions of the output signal that represent signal source V 2 and signal source V 3 on Figure 13-4. Signal source V 2 has a higher frequency than signal source V 3.

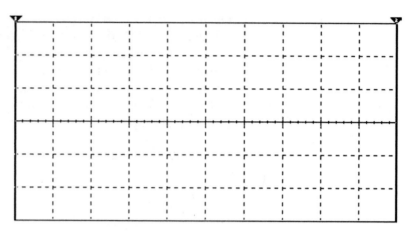

**Figure 13-4**   Scope Display of 74LS157N Multiplexer
Circuit Signals

## ● *Troubleshooting Problems:*

6. Open circuits file **d13-01d**. This circuit has a circuit problem; the multi-

   plexed output looks strange. What is wrong? The problem is _____

   _____

   _____

   _____.

7. Open circuits file **d13-01e**. This circuit is supposed to have an output as

   displayed in Figure 13-5. What is wrong? The problem is _____

   _____

   _____

   _____.

## ● *Circuit Construction:*

8. Open circuits file **d13-01f**. Using the components and ICs on the
   workspace, construct a circuit that will multiplex signals from sources
   V 1, V 2, V 3, and V 4. V 5 and V 6 are the control signals. The circuit is
   partially connected; all you have to do is connect the signal sources.
   Draw the resultant waveform on Figure 13-6.

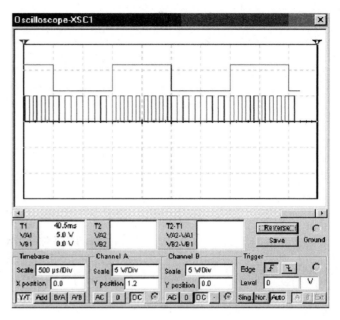

**Figure 13-5**   Proper Scope Display for Circuits File **d13-01e**

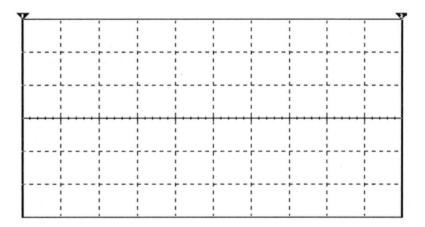

**Figure 13-6**   Multiplexed Waveform for Circuits File **d13-01f**

## Activity 13.2:  Demultiplexing Digital Data

1. A Demultiplexer (DEMUX) circuit receives multiplexed data from a transmission source and converts the data back into its original state consisting of multiple signal sources. These circuits range in complexity from basic combinational circuits to very complex ICs.

2. Open circuits file **d13-02a**. In this circuit, a multiplexed signal is demultiplexed into four separate signals. Activate the circuit, observe the four signals and draw them on Figures 13-7, 13-8, 13-9, and 13-10. Notice that Channel A on the oscilloscope is monitoring the $Q$ output of the flip-flop (U1A). Draw your displays in reference to the $Q$.

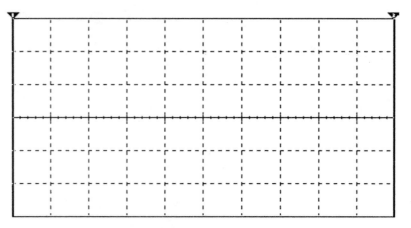

**Figure 13-7**   Demultiplexed Data Output A
from Circuits File **d13-02a**

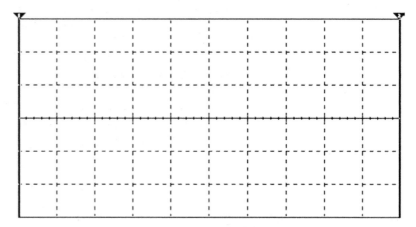

**Figure 13-8**   Demultiplexed Data Output B
from Circuits File **d13-02a**

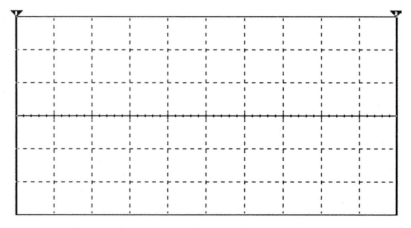

**Figure 13-9**   Demultiplexed Data Output C
from Circuits File **d13-02a**

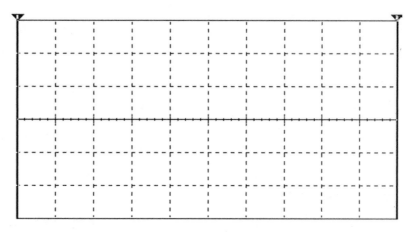

**Figure 13-10**   Demultiplexed Data Output D from
Circuits File **d13-02a**

3. Integrated circuits such as the 74LS138, 74LS139, and 74HC154 are classified as decoder/demultiplexers.

4. Open circuits file **d13-02b**. In this circuit, a 74HC154NT is being used to demultiplex the incoming data (on the ~G1 input). When the proper binary code is placed on the A, B, C, and D inputs, one of the output lines will be activated and the data will go out on that line. Activate the circuit and enter the resultant data in Table 13-1. The object of this exercise is to determine which output has data on it as the input switch settings are changed. The first step places the data on Pin 1.

| Input Switch Settings | | | | Output Pin Number |
|---|---|---|---|---|
| **D** | **C** | **B** | **A** | |
| 0 | 0 | 0 | 0 | 1 |
| 0 | 0 | 0 | 1 | |
| 0 | 0 | 1 | 0 | |
| 0 | 0 | 1 | 1 | |
| 0 | 1 | 0 | 0 | |
| 0 | 1 | 0 | 1 | |
| 0 | 1 | 1 | 0 | |
| 0 | 1 | 1 | 1 | |
| 1 | 0 | 0 | 0 | |
| 1 | 0 | 0 | 1 | |
| 1 | 0 | 1 | 0 | |

**Table 13-1**   Using the 74HC154NT to Demultiplex

*Table continues next page*

| Input Switch Settings | | | | Output Pin Number |
|---|---|---|---|---|
| **D** | **C** | **B** | **A** | |
| 1 | 0 | 1 | 1 | |
| 1 | 1 | 0 | 0 | |
| 1 | 1 | 0 | 1 | |
| 1 | 1 | 1 | 0 | |
| 1 | 1 | 1 | 1 | |

**Table 13-1**    *(continued)*

5. Did you notice that the multiplexed data input to the circuit is on enable line ~G1. For the circuit to work, the other enable line ~G2 has to be placed LOW.

● *Troubleshooting Problems:*

6. Open circuits file **d13-02c**. This is the same circuit as Step 2, but it isn't working properly because of a component problem. What is wrong? The

   problem is that _____

   _____

   _____

   _____ .

7. Open circuits file **d13-02d**. This is the same circuit as Step 4, but it isn't working properly because of a connection problem. What is wrong? The

   problem is that _____

   _____

   _____

   _____ .

● *Circuit Construction:*

8. Open circuits file **d13-02e**. Using the components and ICs on the workspace, construct a circuit that will demultiplex signal data coming from the signal source in a manner similar to Step 2. This circuit only has two outputs, A and B; display the two outputs on the scope face simultaneously. Draw the constructed circuit on Figure 13-11. Draw the scope display on Figure 13-12.

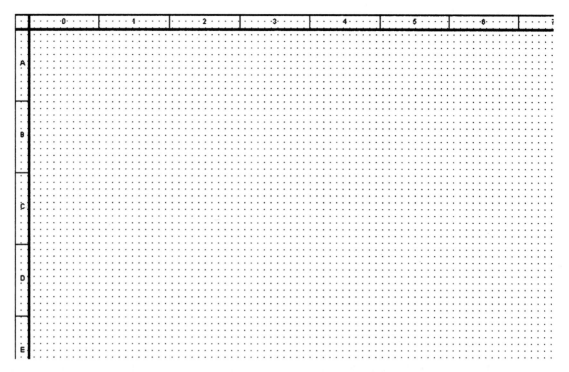

**Figure 13-11** Constructed Demultiplexer Circuit

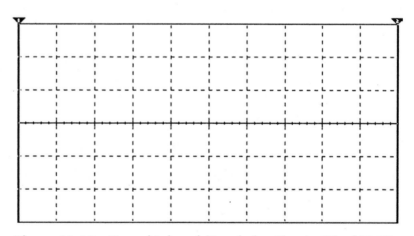

**Figure 13-12** Demultiplexed Signals for Circuits File **d13-02e**